How to build a radio station

Written by:
Dave P Walters
2006

First edition (How to build a radio station) 2006
Copy-edited and typeset by Dave P. Walters in Bromley, Kent.

Published by Dave P. Walters and Lulu

ISBN 978-1-84728-707-6

Contents

Table of figures and diagrams

Note from the author

I have worked in radio for 16 years, half of my life, starting at hospital and student stations before progressing into professional well funded radio. In that time I have been involved in building many stations both very big and very small, many more studio builds and a good few rebuilds. Some have started as empty buildings, others adding extras to stations and even extra stations to existing radio sites.

As my career, developed and my understanding of the industry grew, it became clear to me that few people do what I do. There are many specialists within radio who concentrate on this and that, but few who generalise such as the station engineer who needs to know everything.

This book is an attempt to bridge the skills gap between electronics engineering, bands PA systems and broadcast engineering.

The book is also required reading for aspiring broadcast engineers who are not producers, sound engineers, imaging engineers nor Cool Edit operators. Rather they are skilled system analysts, design engineers, electronics experts and diplomats. This book shows you how to be all of these things, except the diplomat.

This book was written on a brilliant HP-Compaq nx7010 Laptop, on the 0702 Bromley South to London Victoria and the 1804 London Victoria to Bromley South trains during 2005 & 2006. The journey in each direction is 18 minutes.

I drink tea with milk, during daylight hours and if the liver holds out, lager during the darker hours.

Good luck and happy building.

Dave P. Walters 2006

Introduction

Why do you want to build a radio station? It's hard work, it's dirty and more than anything else it has a way of sucking the life out of a person. It's a bit like a child, you create a thing that seems to live and breath, and needs nurturing at every stage of its life.

In the early stages a radio station has to be taught what to do. In the middle stages it has to be loved, fed and watered to keep it ticking along. Towards the end, it has to be cared for like a favourite aunt, who dribbles a little, has forgotten what she was doing, and is surely heading to the grave.

All radio stations die. Therefore before you start creating this life and living and breathing its emotions, pains and tears, you have to ask your self why you want to build a radio station?

Why radio?

Radio has a certain life that TV doesn't. The speed radio can deliver a new message to the public is the time it takes the person with the microphone to read all the words out in the right order. TV tends to be slower as very few of today's TV programmes are actually live. Some current affairs and news programmes are, but they are so heavily formatted with auto-cues, producers and directors that broadcasting new radical information or messages are the last thing on anyone's mind.

Radio is instant, no delay between the idea and the listener, all the presenter has to do is talk! It may not be the best radio, and your career might not progress as you hoped, but it's done. Published, out there, never to be edited, withdrawn, cut or modified.

That's why radio is attractive. It might not be a new message, you might not be able to wait for the colour video on MTV, or for the interactive website to be launched, you might just want to play this brand new home made CD or white label, it's the music of the moment, it's what the kids want, it's what the olds want, or it's what you want.

Alternately, you might be one of the new generations of entrepreneurs who see radio as where the money is after all, if you try hard you can have very low overheads, huge audience, and sell commercial airtime to fill in gaps, both on the air and at the bank. The average modern station should be returning between 35% and 60% on sale, which is a very healthy way to make money.

You have been told to build a radio station

Then you, like 99% of the people in the world who build radio stations, are working for someone else. That's the way it is, and that's the way the world works. In some ways, how great is that. Someone else has found a pile of money, had an idea and has chosen you to create life. That employer has chosen you from thousands of people who would love to do what you're about to do, and now its all yours. They have trusted you with their dream, their investment, and their baby!

You want to build a radio station

Good for you, it's modern and sexy, and you're actually going to do it. It's your idea and your project. This book is not going to help you raise the money, or put the licence application together, but it will guide you through good modern practice, neat ideas and tricks, and help you craft the station you have always dreamed of. This is your baby and you are going to watch it grow from your idea to on air and, if you're doing it right, to a record audiences, not to mention a huge amount of satisfaction.

How do you build a radio station?

Now there is a question and half, and luckily we can now start to guide you through what you need to do, but first you need to know what you are trying to achieve.

What are you trying to achieve?

In your brief from your boss or from your licence application you will almost certainly have some guidance as to what you are trying to achieve. If not, you probably need to ask.

Below you will find three common types of station that fit most requests covering most station formats and target audience demographics. Invariably there are grey areas between all of these, and there will always be examples of new or unusual designs that we haven't directly addressed.

These three examples we will refer to again and again through this book, so it will be worth spending a few minutes familiarising yourself with them.

Example 1. HiRize FM

You have been told by a mate that you are just the person he needs to build his idea for a small scale community radio station, operating out of the spare room in his flat on the 23rd floor of an inner city high-rise. For three hours a day the station will be on-air playing the latest "*banging-toones*". The budget for building this station will be tight.

Example 2. PopFM

You have been told you are building a modern dance music station for a mass market audience aged between 15 and 40 years old, with a slight female bias. Your station will be staffed 24 hours a day with live presenters playing dance and pop tunes all day long. Apart from weekend evenings when pre-recorded programmes featuring guest DJs will be played out. The station is part of a group of stations and there is probably some reasonable money to set this up.

Example 3. NewsTalk Radio

Your station will be a 24 hour speech station. Its target will be intelligent people who have money to spend. They are looking for thought provoking speech based entertainment, with a combination of news, interviews, and phone-ins. The station will be in a mass market and competing against the state broadcaster who has significant resources to throw at similar programmes. You will have reasonable resources for what you are trying to do, but don't have money to waste.

How much will it cost to build?

This is an impossible question to answer, but that's your job now, so let's see what we can do.

There are two approaches to this.

Firstly you could simply ask how much do you have to spend? It's quite likely your boss has an average grasp of finance and so he or she will try and convince you that you can do this job very cheaply. At this stage, your job is to smile in a very non committal way, and go do some research. Never agree to anything, people will hold you to it!

The more likely option is that your brief from the examples above will give you a feel for what you need to do, and if you don't have that yet, keep reading we will get there. From the brief you can go and see some suppliers and tell them what you need and they will put some costs together.

In 2006 these figures would be a rough cost guide based on the station types mentioned above:

HiRize FM, community station in a bedroom, no structural work, just equipment £2000
Small PopFM, on a budget with basic sound-proofing and equipment £100,000
Large PopFM, with several designed studios, equipment and lots of staff £500,000
NewsTalk Radio, with lots of designed studios, equipment and staff £2,000,000

Budget

In order for you to be able to set a good budget to work with you need to make sure you have a good specification. Suppliers who build radio stations, or sell you all the bits to make your own, all want to make a profit but they are not blatant thieves. Contact other radio stations in your area and ask who built their station. They want your business, so whilst there are always deals to be done, concentrate on getting the good specification and stick to it. Once you have this you will find it easier to compare one company's quote against another. If you have three quotes from three companies based on three different specifications, you simply cannot compare like for like and you will not be able to work out who is offering the best value.

Specification

Buildings

Do you have to provide the building or do you already have one to build the station in? Not an uncommon question. It's more likely that your boss will tell you that the two of you are going building hunting and you need to help make a choice.

Things to look out for:

Listed

Avoid listed or state protected buildings, you will find it hard to get approval for any building work that alters the structure. You will probably not be allowed to put aerials or satellite dishes on top of a listed building and generally really old buildings are not suited to today's modern world of cables and pipes between floors and rooms.

If you do end up with a listed or protected building, expect your building works to cost twice as much and take twice as long.

Domestic premises

Smaller radio stations in domestic premises can be ok but only if you are aiming small. Houses are not really suitable for things like large air-conditioning plants, the weight of the studios, or the foot traffic the building will see over the next few years. In other words it will simply wear out.

Commercial premises

Commercial premises are usually the best choice all round. Good solid structures that your studio sound-proofing will depend on. Large empty shells also work as you can add your own solid structure within. Look out for floor to ceiling height, anything less than three metres and you will simply not have enough space to fit your sound-proof floor, ceiling, air conditioning and cable access.

If the building has a number of internal support pillars, make sure they are wide enough apart to avoid one popping up right in the middle of your studio.

Is there enough room to fit your studios, technical areas, desks, meeting rooms, reception, workshops, canteen and kitchens and staff in? Again work with your boss to find out how many staff you intend to have in your station.

There is always a debate between solid concrete floors and computer flooring. They both have advantages. The solid concrete is great for structural work and can have raised computer flooring laid upon it. Computer flooring is great for running all service cables below, but not structurally useful. That can always be resolved by taking up the computer flooring in areas that need structural work, that's assuming that below the computer flooring is concrete.

Floorboards are the worst situation to find yourself in. You need to get your architect and your structural engineer to talk to each other before committing to this property. Your studios will by their nature end up being very heavy, this is what makes the sound-proofing work. Your architect will tell you how heavy the studios will be and your structural engineer will tell you what the maximum floor loading should be. If the floor boards are not strong enough then walk away. Fitting new floor supports and re-engineering the whole building is an expensive job.

If you do go for floor boards you will see 3 to 6 months after the build being completed, that some natural settling of the building work will take place. So expect door frames to move, and steps to become wobbly. If your structural engineer has said its ok it probably will be, but wood is a natural product and it will bend, you need to expect that. Make sure things like your cables have a little slack to accommodate this shifting.

You may find with floor boarded premises that you need to be more sensitive to fire risks. In areas you may have planned to be your kitchen area this could be a problem. It's normally all resolvable, but again care, thought and cost are probably going to be involved.

Regulation: Building, Electrical, Fire, H&S & The Law

Throughout this book techniques and concepts are described that are necessarily regulated by government or law. Buildings need to be constructed well for safety, with Fire and Electrical safety considered at all times.

There is continual reference to consultants, who in the most part should know and understand your legal requirements. In all instances make sure that your work meets the required codes and standards in your area.

I have not attempted to cover the legal aspects here because those requirements are simply too large for what is a technical book, get good local advice and keep within the law. Besides the last thing you want to happen is for your hard work to catch fire, or the building to collapse.

Inner-city or out of town?

The actual location is important. Consider some of the following questions.

- How will people get there?
- Will there be enough parking places for staff and guests?
- Will the staff be all professionals who will earn enough to drive, or will they be younger people who will need to use public transport?
- If access to public transport is required are there near by bus and train stations?
- Will your location deter guests visiting?
- Inner-city usually means higher rent, or building costs and a compromise may be required to reducing studio size or desk space.
- Are there parking spaces for service companies such as cleaners and couriers to carry out their work?
- Are there building restrictions that might prevent you adding aerials or satellite dishes to you site?
- Loading and unloading should be easy.
- Can you easily get rid of waste products, such as cardboard boxes?
- Are rest areas needed?
- Are you providing kitchens or a canteen?
- Are there nearby food outlets or retail areas?
- Is there space for the company to expand?
- Are the local building codes or regulations going to allow your station to operate?
- Can you get access to the building 24 hours a day?

Once you have found answers to all of these questions and points, you are in a strong position to choose the right building in which to place your radio station.

Building Plans

Once a building is chosen then you can plan how the studios and office spaces are going to fit within it.

No matter what your company tells you about how many staff the station will house or employ, as an engineer you have to assume your company is wrong! It will always get bigger.

Immovable physical assets like studios tend to stay where they are once built and rarely grow from one studio into three or four. Desk space though is another matter. The work force will always get bigger. During any period of company success or expansion, someone will make a good case for why they need more staff. The boss may end up wanting a secretary or the sales team might need to expand to get the money through the door.

As an engineer, you need to think ahead to cope with expansion. This will come into play, usually about half way through the build period. Almost certainly with 12 months of the radio station launching.

So whilst the number of studios will probably remain the same, you will end up needing more office space.

Remember to check out the local legal requirements on the minimum amount of space per person within the office. You may get to a point where you simply can't squeeze any more people in.

Departments

In an average size radio station, such as that in example 2, PopFM, it would be reasonable for the following departments to exist. Departments normally like to be grouped together to allow people in the same field to work ideas out as a team. Nobody likes hot-desking, so you could start working on the theory that you will need the following number of desks:

Programming

Programme controller or director.
Head of music.
Programming Assistant.
Presenter(s).
Total desk requirement, 4

News

An average size station may employ one or two news people, who may share a desk.
Total desk requirement, 1

Marketing & Events & PR

Maybe one or two people.
Total desk requirement, 2

Senior Management

The boss and his secretary.
Total desk requirement, 2

Engineering

You.
Total desk requirement, 1

IT support

This will very much depend on the overall size of the company, but in the modern world you should start off assuming at least one person.
Total desk requirement, 1

Accounts

At least one accountant will be needed for any average sized station, maybe two.
Total desk requirement, 2

Sales

If your station is to make money you will need a sales team. If your station only takes syndicated programming and commercials, you might get away with this being one person. A small station selling airtime for it's self will need at least 4 sales people. This is also the area with the most rapid growth and contraction will take place, so make sure you have expansions space.
Total desk requirement, 6

Commercial Traffic/Scheduling

This is the liaison role between sales and programmes, this person places all of the sold commercials into the programme logs and reports back to sales that they have been played, so that invoices can be raised and money made.
Total desk requirement, 1

Reception

This is your company's public front when people walk through the door, it may need to be a separate room to the rest of the station and take up more space, but at the very least you will need one person here.
Total desk requirement, 1

Other

Each company and radio station has its own unique needs not listed here, but other building space will be required for cleaners, store rooms, food to be stored in or cooked. You will need to account for all of this space.

A quick head count of all of your departments means you need to set aside enough space for at least 21 desks.

Studio Facilities

In PopFM, it's been decided that you need two studios and a production area. The first studio will be used as the main on-air studio where all the main programmes will come from. The second studio will be used as a backup on-air studio in case the main one breaks or needs some work. The rest of the time it will be used for production of jingles and commercials, and for pre-recording some of the weekend programmes with guest DJs.

The production area will be used to listen to songs, copy CDs and occasional bits of editing that don't really need a whole studio.

Technical Facilities

In most station set ups it is traditional for the stations main technical resources to be grouped into a Central Technical Area (CTA) or Racks Room. This is where you will find things like the electrical distribution, the telephone exchange, computers and servers, audio processing and links to the transmitters. As an engineer building a radio station, this is going to be the hub of everything that happens, the genuine heart of the beast you are creating.

Drawing up the plans

Now you have considered the scope of the work you need to do and have the basic ideas of what kind of resources you will need to make your station work, you are in a position to make some detailed building plans.

The highlights so far for PopFM are 21 office desks, 2 studios, a production area and a Central Technical Area.

If at this stage you are still looking for buildings it should give you some idea of the space you will need in your building to squeeze it all in.

How much space does a desk take up? Look round your office right now and start measuring an average desk. Check out the legal requirements on the minimum amount of space between desks.

For an average sized studio you will want to set aside at least 4 metres by 4 metres. Don't forget if you are going to build the studios with sound-proofing the walls will be maybe 30cm to 50cm thick.

The size of the technical area will be dictated by how complex your station is going to be. A simple station will probably need at least 2 or 3 head height 19 inch racks of equipment, to house all of the systems and servers you will need. Again you should look at something like 4 metres by 4 metres to make all of this fit!

Drawing the plans

Now you have enough considerations, like how the station will operate and how much space it will take up. You may have a couple of premises in mind and now you need to draw up your plans.

This will help you see if all the things you want will physically fit into the space you have. They will also form one of the first parts of your documentation.

Documentation

All the way through the life of the station from this point onward it's going to grow and change. It will in its early days have several contractors working on site. To attempt to keep all of the plans in your head over the build period, and describe them to anyone coming to work for you is simply folly. There is just too much to remember. You need to write it down.

At this stage the documentation will probably still be very basic, and that's ok. A drawing of your proposed building lay-out will be fine.

This is what you take to your architect, or the company you will employ to physical do the station build for you. They will want to get a feel for your ideas. Paper is the best place to start. There will almost certainly be several site visits to see how the plans compare to the real space. With your architect, or building company, those plans will grow and be modified to precisely define what will happen.

Scale drawings

Scale drawings will help people better understand the requirements. The diagram below is an example of what your first stage plans might look like...

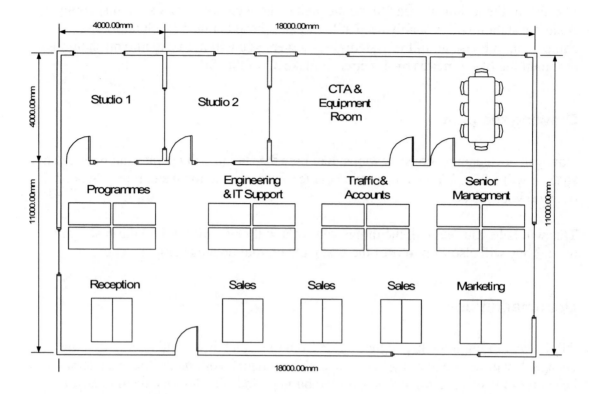

Figure 1 PopFM station layout

Operational Requirements

At this stage you have a good idea of the building you will be occupying; you have a fair idea of what the station will be trying to achieve in terms of business and the audience you will be trying to attract. You know how many studios you intend to build and roughly how they will be used. Now is the time to think about how those studios will be put together to achieve that.

The following details will be concentrating on equipment needed purely for the basics of broadcasting and not looking at building infrastructure, such as power or air-conditioning.

Station Types

In our three examples, we cover the basic types of station available, bedroom radio, a small commercial music station, and a larger talk based station. Each of these will require a different technical configuration in order to optimise the way they will be used. In many cases the design choices at this point will remain in place for a long time, typically ten years or so. Now is the time to make the right choices. Consult with similar stations about their designs and configurations. Determine what works well for them and what they may have done differently given the benefit of hindsight.

Once those choices have been identified, you will need to look at the specific types of equipment, and supplier who can best deliver your needs.

Domestic versus Professional Equipment

The cost difference between professional and domestic studio kit is significant, maybe by a factor of 10.

A domestic CD player from a local high street store may cost £100, although the professional version may cost £1000. Professional equipment is designed to protect itself from its users. The buttons to operate it are larger and heavier and it is better ventilated. It is assumed the professional user will attempt to destroy the equipment on an almost daily basis. The domestic user has worked hard to earn the £100 and wishes to treat their investment with respect, and will press the buttons gently and in a caring way. Not so the professional user.

There are also some technical differences. In most cases the professional equipment will sound better. These days genuine sound quality does not seem to be a large factor by anyone's measure, that coupled with the explosion in digital technology can often mean that the performance of domestic equipment is at the very least comparable to that of professional.

The professional equipment will also offer advantages such as Balanced Audio inputs and outputs, compared to only unbalanced on domestic, or the standard professional AES digital interfaces.

For HiRize FM it may be best to supply and use domestic equipment. The usage will be low, which means domestic equipment will probably survive for at least a few years. There is also an argument that in a low budget station that the costs of replacing a damaged domestic machine at say £100 is much lower than the cost of repairing a professional machine at maybe £200. All round it's a better choice.

The only other consideration is that domestic equipment tends to be unbalanced, which in a small station is not normally an issue. Assuming that the small scale station means the budget is tight and so spending big money on a professional CD player it is unlikely as it may end up costing about half of what you could spend to build the entire radio station.

In a larger station such as PopFM and NewsTalk Radio, professional equipment is essential to the technical success of the project. In a modern office or broadcast environment there are many hundred of electronic signals. Domestic equipment will suffer interference from these electronic signals which will result in your transmission output having noise and distortion all over it.

Professional equipment is designed to shield itself from electronic noise, and to transmit its signal in a way that is highly immune to noise. This will result in a very clean station sound. Unbalanced domestic equipment in a large technical environment will suffer from buzzing, hums, pops and crackles.

Example 1, HiRize FM

This station is going to be low cost, based in someone's house and will only be on air a few hours a week. Its purpose will be to play the latest pop records to the kids.

You probably have the feeling that your friend doesn't want to spend too much money, and that the spare room in his house may also get used for occasional Sunday lunches and watching TV.

Equipment you will need.
2 CD players
2 Turntables
2 Microphones
1 Mixing desk
1 amplifier
1 pair of speakers
1 set of headphones
1 Transmitter
Interconnecting cables

This is the basic list of items you will want to use, and will result in a studio facility providing the very basics required for radio.

Example 2, PopFM

This is the more professional end of the music market and you will have some money to spend, although the size of the organisation and the complexity of the aims on air will dictate the eventual size and budget for what you are trying to put together.

Equipment list

Studio 1 & 2

Professional 8 to 12 channel broadcast audio mixer
2 Broadcast quality microphones
2 Professional CD players
Minidisk player/recorder
DAT machine
Amplifier
Set of speakers
2 Pairs of headphones
Distribution amplifiers
Outside source selectors
TV
Email and internet PC
Interconnecting cables
Computer play-out terminal
Telephone phone-in system and TBU interface unit
Mains and power distribution and management

Production Area

Small audio dubbing mixer
CD player
Amplifier
Set of speakers
Minidisk recorder/player
DAT recorder/player
Interconnecting cables
Computer play-out terminal
Mains and power distribution and management

Central Technical Area

At least two or three 19inch rack-mount cabinets
Significant amounts of interconnect cables from studios to the CTA
Mains and power distribution and management
Distribution amplifiers
Patch panel
Monitoring station
Amplifier and Speakers
Studio switching system
Computer play-out system
Servers, IT hubs and switches
Backup computer systems
Studio telephone system base station/server
Studio ISDN equipment
Off air receiver for monitoring
Off air logging of monitoring for legal purposes
Office servers
Office telephone system
Office telephone lines
News provider satellite receiving system
Internet or ADSL lines
Office IT network Hubs and Switches
Selection of spares and tools for maintenance and repairs
UPS backup power systems
Links to transmission system
Backup links to transmission system
If your station is housing the transmitter on your premises, then of course you will need a transmitter

Office space

21 desks
21 chairs
21 telephones
21 computers

Office infrastructure

Each desk position will require at least two power points and one of these will usually have a 4-way extension board hanging off it.
Each desk position should be provided with 4 CAT5 points; these will service the office PC and phone per desk, as well as any extra items such as printers and fax machines.
The building should be fitted with heating and air-conditioning.

The design of the office space during the build should be considered carefully. No matter how good the design, within a month of the office being used by people it will have to change. The better the planning the fewer changes will need to be made. Employing a good flexible design will make those changes relatively painless. A rigid design will make the work hard to achieve and result in extra cost and difficulty for all involved.

Again Health and Safety and legal requirements will have to be adhered to at all times.

Example 3, NewsTalk Radio

This type of station can be much more complicated. As in all the other cases and examples the level of complexity will be dictated by finding out what the staff in the station expect and what you have in terms of budget.

A talk based station could actually be very simple. All you theoretically need is one microphone, a telephone (with studio interface), the mixer and the transmitter. As it's a talk station, there's no need for music so there is even less kit to think about. The reality is that it's going to be at least as complex as the PopFM example and probably more so. The PopFM design could work quite well, but it will depend on what the programming team expect.

If the station expects its NewsTalk presenters to operate all their own equipment whilst talking on the radio then PopFM is for you. If however, the station is going to be very heavily produced and have assistants and producers all over the place, you need to look at bigger resources.

If your breakfast presenter is going to interview guests, talk to people on the phone, and cross to news and travel, without ever pressing a button himself, then you might consider housing the presenter in a studio, and the production team in a control room.

In this case the presenter should have a studio to themselves, which contains just microphones, headphones, TV and PC. Then in order to operate the station, you will need a control room, with a large window for a perfect view of the studio. A Studio Manager will operate the mixer. You will have a Producer who directs and produces the programme. They will also meet and prep the guests. You may also have a Phone Operator who answers hundreds of calls per hour and only when they get the best callers transfer them over to the studio to talk with the presenter. There may also be a Runner to collect guests from reception and them to the control room, and also gets the coffee.

In essence you are looking at a bulked up version of PopFM, which also has a voice booth for the presenter to sit in.

This time, however, it simply needs to be bigger; already in the control room we have identified several key roles, studio manager, producer, phone operator and runner. This means that even if no extra kit is required, you definitely need more space.

The following diagram is an example of a typical NewsTalk studio layout; again your local circumstances and requirements will dictate your final layout. I have also ignored the office staff question in this example for a couple of reasons.

Firstly, if you have a large production team per programme, they will need to all sit together whilst preparing for the programme. The station will need to decide how many programmes each day there will be, for example; early breakfast, breakfast, midmorning, lunch, afternoon, drive, evening & overnight. This is lot of programmes for a normal station and it would be more common for many of these to be merged, such as midmorning, lunch and afternoon into one long "day time" programme. This will be more cost effective for the programming department and also mean you will need fewer desks.

In this example we have identified 8 separate programmes. Assuming a team of 4 per programme, then that's 32 staff you need to provide for, not to mention any extra weekend programmes or staff. Maybe you are planning to hot-desk and get the night staff to use the day time desks??

Secondly, if you are starting to put together office plans for a company where just one department is 32 people, you need to look at getting some help. This could be the senior management team, who will all have opinions on how the office layout should be, or maybe you could get help from your company facilities manager, if there is one.

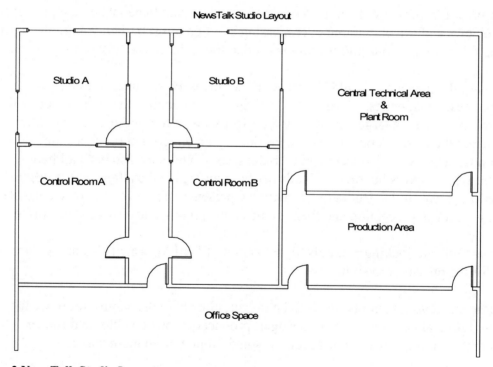

Figure 2 NewsTalk Studio Layout

Equipment List for NewsTalk Radio

The size of this station suggests that the Control Room A and Studio A set up should be the same specification at Control Room B and Studio B. This differs from PopFM, where studio 1 was a slightly higher specification.

This is because speech radio is much more likely to pre-record long programmes that need the same level of technical complexity as the main on air studio. It is also a good idea to keep studios at the same level of specification, as it allows programmes to move from A to B in the event of a fault or routine maintenance.

The equipment list for the Control Rooms is very similar to that of PopFM, although the layout will be slightly different. Remember also that the Studio will be operated via the Control Room, and so there needs to be a clear communication system between the studio and the control room.

Control Room

Professional 8 to 12 channel broadcast audio mixer
2 Broadcast quality microphones
2 Professional CD players
Minidisk player/recorder
DAT machine
Amplifier
Set of speakers
2 Pairs of headphones
Distribution amplifiers
Outside source selectors
TV
Email and internet PC
Interconnecting cables
Computer play-out terminal
2 Telephone phone in system and TBU interface unit
Talkback system between studio and control room
Mains and power distribution and management

Studio

4 Broadcast quality microphones
4 pairs of headphones
Set of speakers
Amplifier
TV

Email and internet PC
Computer play-out terminal (possibly a copy of the Control Room system)
Talkback system between studio and control room
Mains and power distribution and management

Production Area

CD player
Amplifier
Set of speakers
Minidisk recorder/player
DAT recorder/player
Computer play-out terminal
Talkback system between studios and control rooms
Mains and power distribution and management

Central Technical Area

At least 5 19-inch rack mount cabinets
Significant amounts of interconnect cables from studios and control rooms to the CTA
Distribution amplifiers
Patch panel
Monitoring station
Amplifier and Speakers
Studio switching system
Links to transmission system
Backup links to transmission system
Computer play-out system servers and IT hubs and switches
Backup computer systems
Studio telephone system base station/server
Studio ISDN equipment
Off air receiver for monitoring
Off air logging of monitoring for legal purposes
Office servers
Office telephone system
Office telephone lines
News provider satellite receiving system
Internet or ADSL lines
Office IT network Hubs and Switches
Selection of spares and tools for maintenance and repairs
UPS backup power systems
If your station is housing the transmitter on your premises, then of course you will need a transmitter

Office space

This will be a large office and will require significant planning, based on the number of programmes, team sizes, departments and crucially the opinions of senior management.

Office infrastructure

Each desk position will require at least two power points, and one of these will usually have a 4-way extension board.
Each desk position should be provided with 4 CAT5 points; these will service the office PC and phone per desk, as well as any extra items such as printers and fax machines.
The building should be fitted with heating and air-condition.

Open Plan or Offices

There is much debate among modern business leaders as to which is best, an open plan office space or lots of smaller closed off offices.

The advantages of open plan are considerable for the engineer at design stage, and usually much easier to build. Modern office buildings tend to use computer flooring which allows cables to be run below it. The flooring itself consists of removable tiles supported on little pillars about 15cm off the main concrete sub-floor. This means if you provide power and data points to a desk position and someone wishes to move their desk a few metres, you can usually move the floor tile that holds all those cables to the new desk position without too much trouble. This will save considerable time and money. The open plan office design will also cut down on your costs in terms of building walls and will allow an office wide air conditioning system to do its work, rather than several small systems spread around several offices. Open plan is also very space efficient. Whilst a wall is not usually very thick in an office space, the space around it that becomes unusable is considerable.

In environmental terms the open plan office allows for teams to work together and makes simple point to point communication within the office very easy. It can cut down on the growing problem of someone emailing a question to the person in next office, and make them talk to each other.

The open plan office also suggests that the team members are all of an equal status. If half the team has an office and the other half doesn't, it will be very easy for a 'them and us' culture to build up.

The disadvantages of open plan, tend to be in terms of privacy. It will be very difficult for managers to hold small team meetings, or discuss sensitive matters such as business deals or staff problems. This can be resolved by the provision of several meeting rooms of various sizes. A few large board room style meeting rooms, suitable for both staff and external clients, as well as several smaller rooms, suitable for one or two people to carry out staff appraisals or interviews.

For the engineer, the multi-office concept can be more of a challenge. Each room may require its own light switches, power distribution, air conditioning and HiFi system. All of this will increase build time and costs.

Creating the specification

You should now be in a position where you know what kind of station you will be building. How many staff it will have and what roles they will be filling. You should know how many studios you need to provide, and what those studios will be used for.

Jot all of this down, or preferably make this available in some kind of digital format such as an MS Word document. You now need to decide if you will be building this station yourself or contracting a single team of specialists, or several smaller teams, who each specialise in key things such as studio builds and office builds. Either way your documentation will be the key starting point to them understanding what you are trying to achieve.

Should you do the work yourself? If the station is small enough and you are skilled enough, then this is very reasonable - if you have the time.

As an idea of what is involved, a small station build, such as PopFM, 2 studios, a CTA and 21 office desk, the work may follow this kind of plan.

Month 1
10 builders install studio shell and sound-proofing. Assemble any office partitions.

Month 2
5 Engineers, installing the infrastructure to the station, office data cabling, studio interconnect cabling.

Month 3
10 tradesmen provide the finishing touches to the build, such as electrical power, doors, carpets and air conditioning.

Month 4
5 Engineers install and connect studio equipment and testing. Install office phone systems and computer systems.

Month 5
4 Engineers provide training to staff and users. Assist with dry running of programmes. Fix bugs in the system and finally put the station On-Air.

Month 6
2 Engineers providing support for training and bug fixing, whilst the station is on air. This is also the time to go back to the original contracts and specifications and check that the installation company has provided all that you were quoted for.

How to get someone else to build the station

This is the point where you might be feeling overwhelmed and be looking to contract out. Indeed this is how 95% of all radio stations are built. The details of floor loading, acoustic sealing and damping are quite involved, and can take years to learn. If you don't have that time then get someone else. If you do have the time, we will now take you through the basics of construction from the ground up.

How to find contractors and specialist installation companies

If you are intending to find specialists to help you build your station, you should now be in the position to pass on your specification. The key question is how do you get the right people to do the work for you?

There are a number of good ways.

Visit Trade Shows

There are a number of broadcast trade shows each year around the world, all of them are jam packed full of companies who will sell you microphones to mixers, and install the whole lot in an empty building for you if you like. This is the place to go to find contacts. There are a number of key trade shows
SBES, in the UK
AES, in Europe
IBC in Amsterdam
NAB in Vegas

This is where you will find the "movers and shakers" of radio from around the world. There are of course many more shows, but these are a good place to start, assuming you have the time and money. If not, then check out their web sites that will normally give a list of exhibiters from previous years so that you can track them down.

Local Contacts

Radio is a small industry and asking around will always brings up names that someone has met along the way. Give them a call, what's the worst that can happen? Failing that; cold-call radio stations you know already exist and ask to talk to their engineers. Radio engineers are a good sort and will be excited you have bothered to call them and will be keen to pass on any useful thoughts or ideas they have. They may even give you names of people they would recommend, or those to avoid!

The Internet

The internet knows everything! Find yourself a search engine and type in your key words such as, 'Build Install Design Radio Station'. Often using the local versions of search engines such as google.co.uk (rather than the international google.com) will give more localised results.

What next?

Contact a few of the suggested companies, email or post your specification to them, then wait and see what happens. The companies who can do what you need will be all over you like a rash. They will be on the phone, wanting to visit you and offering advice. You will get a feel for who is genuine and who is not. The ones that visit you probably mean business. They want to see for themselves what the work entails before giving you a quote. Anyone who quotes blindly without seeing a project should be treated with caution. They will then ask you lots more questions, suggest changes based on their experience and help you feel comfortable and confident with them as a company. If it doesn't feel right, don't do it, get another company. It's your money and your radio station. The company will be working for you. You should at least enjoy it. If you don't think you will enjoy working with this company, try another one.

Self Build

Less scary than it sounds, but still hard work. Rather than contracting a company to project manage the whole build, you have decided to have smaller contractors tackle the individual jobs. This is a valid way to build a station and we will be looking at what you need to know in the next section,.

The Project Plan

You know what you want to build; you may have found the contractor. You now need to get quotes, check they match what you need, maybe pass on your best quote to another provider and get them to quote against it.

Next you need the contractor to come back to you with a detailed plan of what and when they intend to do and how much it will cost.

Take the best quotes back to your company and check that the management agree with your plans. If everyone is happy sign the contract and start the build.

You have your specification, you've found a contractor and they have re-written the specification in a more technical form of which you still approve. You have a cost and a timescale and you are ready to go.

Your job now is to supervise the company doing the build. Make sure they stick to their timetables and ask questions if they start missing deadlines or details. Observe how the station is being put together. If you ever build a second station you may wish to build it yourself, and watching the professionals at work will give you lots of tips for next time, just like this book!

It's important to observe what they are doing so you know how to fix or expand your new radio station. You will need to know, for example, where the power cables enter each studio. This means when you hang your company logo on the studio wall, you won't drill into the bit of the wall with the power in, although of course, you will not be puncturing your very expensive sound-proof box with screws… we hope.

What could go wrong?

The most common problems are attention to detail in the build process, and project slippage, where the completion date continually moves.

In the contract with the construction company you should work out if you want to pay for the job by time, or a single fee on completion. There are advantages to both; if you pay based on time, the guys may finish early and save you some money - although this never happens. If you pay by the job, you protect yourself financially if the job does over-run. You will not incur any extra cost. It also helps focus the minds of the contractors - they themselves are now losing money as they are prevented from working on other jobs.

To help hit the completion date you could build in penalty clauses in the contract. Companies hate penalty causes and may not agree although every loves a bonus and this may yield better results.

Retention of payment

It is reasonably standard for a contracted company to ask for maybe 50% of total payment some time before the work starts, and then the rest upon completion. It is also reasonable of you to retain maybe 10% or 20% of the final payment, until all outstanding issues have been resolved. You may for example have a problem with your air-conditioning that needs resolving before you pay the final figure. It gives you significant leverage with a company who have left the site, and now simply wants paying and to move onto the next job.

If you are working for a reasonably large company you may have company accountants who specialise in contracts and can work this out for you.

Many companies are entirely reputable, and where they have done a fair job to a good standard, they should be paid promptly and with grace. Many companies these days like to delay payments by weeks or months, in order to maximise their cash flow, but this simply destroys good honest people who have worked hard. If someone has done a good job for you, make sure you pay them. Although it may be your company that is delaying any payment, it's your name on the contract and your reputation your company is damaging.

Warranty Period

Just like buying a car or washing machine, your brand new radio station will come with a 12 month warranty. Or at least it should. If you have paid people to do a job, such as put a door and wall here, you probably don't have any warranty. If you have a paid a company to build one radio station, you probably do! Make sure you check the contract before you sign and talk about what this means.

This doesn't protect you from going off-air, but it does ensure that if after 12 months the air conditioning simply doesn't work, or that a door or window has fallen out due to bad workmanship or poor materials, then the company should come back and repair those issues.

At about 10 months into the life of the station, walk around and make a list of all the issues that you think should be covered under warranty, and let your contracting company know what they are. They will probably be expecting your call, because sometimes things just break. Engineers have jobs because sometimes things just break.

Documentation

The project is never complete until you have received the documentation. This is critical for you as an engineer. No one can ever remember how every system or machine works, and you need that documentation. More than that, it will tell you how your station is constructed, where the wires and the pipes are and how they connect together. This will be your bible to fixing this station for the rest of its life. If you leave and a new engineer takes over, how will they possibly know how the station is constructed? They need the documentation.

The primary build documentation should live in a number of folders:

1. Physical Construction. Details of how the studio and office build took place with architects drawings.
2. Electrical and Mechanical. Details of where the power points and light fittings go and how the power is distributed. How the air conditioning and heating systems work.
3. Technical details of the radio station systems. How all the audio devices are connected together, where all the cable runs are. The manuals for all of the equipment. The acceptance and test documentation showing all the systems work as specified and designed.
4. Certification, any documents or certificates required to meet legal requirements for safety and construction.

Never release the final payment until you have the documentation. You will regret it every day for the rest of your life.

Synopsis

You have understood your station's needs in terms of market, style and size.
You have designed the studio and office layouts.
You have found a company to do the build for you, within a budget and time scale.
You have managed the building and installation companies and tested all the systems, and received the final documentation.

How to build your own station by hand, yourself

Do you really have the time and the skill?

This is a really important question. If you genuinely intend to go down to the local hardware shop, get your own supplies and construct everything yourself - then you really are facing a difficult job, especially if you are building anything more than HiRize FM.

If you intend to build PopFM totally alone, you either need help, or plenty of time to do it. To build a reasonable sized station alone is very hard work.

If you have a team of builders and a team of engineers, and would like to direct them with your own plans, then with good project management skills and careful attention a radio station will be born.

How to build a radio studio

Specification

Specifications are critical. It does not matter if this is a one man job or a job for a team of builders. You need to know what you want to achieve. Without a specification the project will lack drive, and will drift as people disagree on what they are trying to achieve. There is no reason to stay with what you later find to be a bad specification, but you must at least start with one.

As with all projects, there are many variables and the range of possibilities are endless, so to help us narrow down our criteria, we are going to concentrate on PopFM.

We know that PopFM will need two studios that will be more or less identical. This means we have one specification, needed for each of the studios.

In a pop station, complete sound isolation (totally quiet studios) will not be necessary, but good sound isolation will be required.

In the modern world of pop radio, however, even poor sound isolation is often acceptable. Many stations choose to process the sound of their microphones very heavily and use very close microphone techniques in order to get the sound they want, reducing the need for perfect isolation.

We will aim high, however, with good isolation! The studio will have a fast reverb with a rapid decay. This will add a little life to the sound of the room, avoiding the dead sound of an anechoic chamber, and also avoiding anything like a lively sound. You need to avoid a studio that sounds like a bathroom. We will come onto this later when we talk about sound conditioning and colouration.

Right now, for PopFM, all we need to know is: Good Isolation maybe 40dB, rectangular box 4m by 4m, air-conditioning, doors, windows, access for power and audio cables.

Why do I need isolation and how much do I need?

If you are trying to talk into a microphone on the radio and tell people the name of the song that you have just played, and suddenly a friend shouts across the office telling you there is a phone call for you, you may find it a little distracting. The listener at home may not appreciate the extra information they are getting about your phone calls, they just want to here about the song!

Indeed, if you are the friend at in the office, you might find it impossible to concentrate on your work when all you can hear is the radio presenter playing records at the highest volume.

A studio with some kind of sound isolation or sound-proofing is the way to go. This will protect the listener and the radio presenter from the unnecessary real world, and will also offer some sanity to the office staff in the radio station.

Also if you have more than one studio then you will need to protect the two presenters and studios from each other.

More traditional arguments about sound-proof studios are about creating spaces where you can craft or work on sound. In order to do that, it would be ideal if the only sounds you can hear are the ones you are creating, and not environmental noises which will make your job more complicated.

How much isolation you need is a difficult question. The answer will depend on how noisy the local environment is, and therefore how much noise you need to keep out. It may also depend on how much noise you intend to generate inside the studio, and how upset the neighbours would be if it leaked out and kept them awake all night long.

Being able to measure and quantify sound is the key. If you are an expert, then you will need to get a very expensive Sound Pressure Level (SPL) Meter and make some absolute measurements of the local environment. Alternately, here is an instant guide

- Jet air craft taking off, 120dB
- Loudest Pop Concert you have ever been to, 105dB
- Busy high street road traffic noise, 75dB
- Busy office with phones ringing and people talking, 65dB
- Quiet Office, 50dB
- Opera House/Symphony Hall, 30dB
- Radio Studio, 25dB

So how much isolation you need will depend on you local environment. If you want to build a radio studio, just in front of the stage at the loudest pop concert, you will need to reduce the local sound levels from 105dB to 25dB, achieving 80dB of isolation, or sound reduction. That is very hard.

If you are basing your radio station in a busy office, then you need to reduce the absolute office noise measured at 65dB to the required noise level for a radio studio of 25dB, meaning a reduction of 40dB. This is also hard, but very achievable.

If you are looking for a genuine scientific and certifiable studio construction then you really need to approach specialists in studio design, noise control, and construction. For the basic guidelines, read on.

The Theory of Sound

In order for you to specify well, bluff your way in studio design, or even actually build your own studio you need to understand the basic physics of sound and how it works.

There are many people who are experts in sound, who have spent years studying it academically, or professionally. What you will learn here is basic, but critical if you stand any chance of achieving a good sound-proof studio.

Sound, at its simplest level, is a vibration of the molecules in the air that you can hear. In essence a musical instrument, or voice or loud speaker all vibrate, which in turn cause the air to vibrate. The air vibrates your microphone diaphragm or the cilia in your inner ear. All of which can be measured to create a sense of sound.

Water is a very good way to illustrate the behaviour of air, and more specifically sound travelling through the air.

Throwing a stone into a flat still pond will result in a nice circular ripple (wave). The wave itself is a disturbance of the otherwise calm water molecules caused by the pressure of the stone upon the water. The pressure causes the water to move sideways to let the stone into the water. As the water moves sideways it applies a force to the water molecules next to it.

A large circular wave will spread around the stone, getting further and further away. As it gets further away its magnitude or height of the wave from top to bottom, gets smaller and smaller.

This is a simple but very effective analogy to keep in mind throughout this section. As any item creates a vibration, it will cause the air next to it to vibrate, pushing waves of energy through the air and eventually reaching you.

Sound primarily travels through air, however, it very happily travels through any medium. Sound will happily travel through solids and liquids, not just air!

Creating Isolation

- Prevent any air from getting into or out of your studio. Air carries vibrations. Create an air sealed box and no vibrations can enter through the air.
- Make sure your studio is not physically connected to the outside world. Practically impossible, but with compromises very effective.

An air sealed box is really quite practical (apart from suffocation) and is the starting point for controlling the sound levels in your studio. The compromise will be the air conditioning as people have to breathe. Attention to detail really can create a room with no holes through which sound can vibrate the air and allow sound to enter or leave your studio.

Physically disconnecting the studio from the world is nearly possible. An average room you has six surfaces, four walls, the floor and ceiling. Normal each one of these surfaces is connected to the outside world and each of those surfaces will pass structural noise from the other side into the studio.

By creating a room within a room, and only allowing the floor of the inner room to come into contact with the outer room, you have removed five sixths of the problem.

The inner room is totally disconnected from the outer. The rooms are totally airtight, stopping air leaks transmitting sound in or out of your studio and will also stop any structural noise transferring structural vibration from one room to the next. There are still many problems to overcome, but we will go through those, and the results will be quite impressive!

Sit the new inner room within a room on neoprene pads. These are a structurally sound in that they will hold the weight of your new room and they will also absorb and not transmit structural sound. Therefore any vibrations that would normally pass noise from one room to the next are mostly removed.

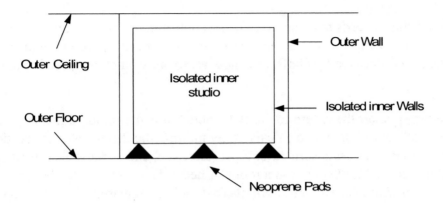

Figure 3 Room within a room

You must have air-conditioning, people need to breathe, but anything that pushes or vibrates air makes a noise and that must be controlled. It must also be connected to the outside with some kind of pipe or duct in order to draw in fresh air and remove stale air.

These big holes in your studio are a problem and are addressed by baffles to trap airborne sounds and rubber couplers to structurally disconnect inside from out, as well as absorbing any direct noise that gets into the system from the environment. If you have ever talked into a long length of drain pipe you will know long pipes transmit sound just fine.

There are plenty of other sources of noise leakage such as doors, windows and cable access, but they can all be managed.

Sound Conditioning

Anyone who has stood in the shower or lay in the bath has probably had a crack at singing. Even the worst voice in the world will sound better in a bathroom. The room will be full of hard flat surfaces, normally ceramic tiles. Mirrors reflect light. Tiles reflect sound.

Imagine an air molecule to be a football, kick it hard at a wall, the ball bounces off. This is because very little of the footballs energy is absorbed by the wall. The ball compresses under its own weight and motion, until the ball has decelerated to zero, then the deformed elastic nature of the ball causes it to spring back in some direction similar to which it came from. This is what happens to sound when it hits a hard flat surface. Little is absorbed and most bounces back.

A similar experience is noticed in caves, an echo of any noise made comes floating back to you. Echo is the term given to a single bounce, and reverb is the name given to multiple bounces, such as those that repeat and then decay, like in a cave.

Reverb in your studio needs to be controlled. To control it you need to be able to measure it. Reverb Time (known as RT60) is the measurement we need and refers to the point where the noise has decayed by 60dB, or is now reasonably quiet compared to the original.

RT60 is the time it takes the reverb to decay by 60dB. It is the standard that acoustic engineers will talk about. It can be calculated by multiplying the acoustic absorption coefficient, between 0 (least absorbent or most reflective) and 1 (most absorbent or least reflective), of each surface material in a room. A heavy fabric known to absorb sound could have an absorption coefficient of 0.7Sa (Sabine's), compared to a solid concrete or brick surface which may have an absorption co-efficient of 0.1Sa When looking for building materials look out for these coefficients.

If you are looking to create a reverb time sounding like a nice church, the kind that choir might like, you probably want to create a reverb time of maybe 2 seconds. For a small space like a radio studio, you want to be looking for half a second.

For radio make your studio sound absorbent rather than sound reflective. Fabric surfaces such as carpets and Hessian and foam wall coverings will achieve this with good results.

Sound Conditioning is the skill of the acoustic engineer to create the type of sound in the studio that you want to hear. You probably don't want a long reverb or a room so dead it is slightly painful. You want a room with just a little life (reverb), but not much more. The skill comes down to structural choices, absorption and reflection choices and also to the positioning of equipment and people within the studio space.

Standing Waves

Most things in the world have a natural harmonic frequency. If you pluck a guitar string, the sound you hear is the result of the resonant frequency of that string. By making that string longer or shorter by holding down one end of the string on the fret board, the resonant frequency changes. It's the same as in a church pipe organ. You blow air into one end of a pipe when you press the organ key, and a note is generated. Change the length of the pipe, or more simply press a different key and a different frequency note is generated. This is the principle of standing waves.

If you have a studio with two parallel surfaces, for instance the near and far wall, you can calculate the length between those walls, maybe 4 metres.

Sound can be seen as a series of waves, the same waves as the ones in the water that you can see when you throw rock in. Those waves have a length. If the length of your wave (l) matches that of the length of your room, they will be naturally in tune.

This means that if the peak of your wave is in the middle of the room, the troughs of your waves will hit the edges of your room and be reflected back. This will cause a new reflected peak will occur in the middle of the room.

The new peak at the centre of the room will add to the first peak, where the noise was generated, and a standing wave will occur, making it louder and persistent. It will not easily be absorbed and fade away

Maths helps us work out these lengths and frequencies:

The formula to use is $c=fl$
c is the speed of the wave
f is the frequency of the sound
l is the wavelength that will be produced.

The speed of sound, at sea level atmospheric pressure, is measured as 330 metres per second.

If the studio has a length of 4 metres between two opposing walls, we can calculate the frequency of wave that will reflect and superimpose itself on the previous wave perfectly to create a standing wave, or a resonant frequency.

$c=fl$, we know c, we know l, so we need to change the formula using high school algebra to $f=c/l$
$F= 330/4$
The resonant frequency of the room will be 82.5Hz

This is a low frequency, so we can expect our room to sound louder at the low base end of the sound range.

The working height of a studio, to the acoustic ceiling rather than any void above it may be 2.5metres.

Using $c=fl$, we know that the floor to ceiling resonant frequency will be 132Hz.

In our basic studio for PopFM we said the room size would be 4m by 4m, and a height of 3m(2.5 after deductions).

This gives us 2 sets of apposing surfaces which will resonate at 82.5Hz, and one set of surfaces which will resonate at 132Hz.

You may be happy with that, but the idea in studio design is to create a neutral room which will sound "flat" across all frequencies. You want your ears to hear the intended original sound, and for the room not to act as its own graphic equaliser to change the sound before you even get to hear it. Or in a recording studio, not to change the way the voice or instrument sounds before you record it. Using sound absorbing materials on the inner walls of the studios will help to minimise these effects.

Room shapes

The problem with the 4m x 4m room is that you have two sets of surfaces set to resonate at the same frequency, doubling the problem.

If you change the room size to say 4m by 3.5m then you would created a new resonant frequency at 94Hz, but at least this would not be adding to the existing 82.5Hz problem.

The next best way to deal with the standing wave resonant frequency issue, is to change the shape of your room. Sound will reflect between two parallel surfaces and have a resonant frequency. What if they are not parallel? Radical as this sounds, it is a very common studio practice.

In our original studio design for PopFM we suggested a 4m x 4m studio design. Making the walls all slightly different lengths would stop parallel surfaces causing standing reflections.

Certainly if you have worked with builders before, you will know how difficult it is to get them to build walls straight and parallel, imagine how confused they will be when you ask them to specifically not do it!!

We are not talking huge amounts here, just a few tens of centimetres here and there.

How does 4m x 3.9m x 3.8 x 3.7 sounds. It sounds a bit freaky. But you could go for a something a little more manageable, which might not result in you wanting to be ill when you walked into the room, Maybe 4m x 3.75m x 3.75m x 3.5m. This way you still have two parallel walls, but you have halved the problem.

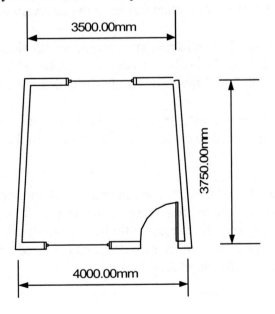

Figure 4 Non liner room shapes

The size of the room will dictate what angles you can get away with. You are not looking to create an odd shaped room, simply take the parallel walls away. That will result in a space you can manage in terms of how it sounds. If you have a much larger studio then a small angular change away from parallel walls probably won't be even noticed.

There are of course curved options to create your studio space, but remember back to your school physics lessons, if you create a nice concave shape it will have a focal point, which is a significant point that sound will all reflect back to. This can be ok, as long as the microphones or presenters are not based in the focal point.

Foundations

Ideally the area you are building studios in has a solid and flat concrete floor. This will be high load bearing and also have some significant mass, which is ideal absorbing sound. If its not you need to assess the floor and work out if it's solid enough to build on. A structural engineer will help you here.

The room within a room principle is going work by creating the outer room, and then building a second room within it. The second room is going to sit on the floor of the first (or outer) room. To achieve as much isolation as possible, you want the new inner floor to connect to the outer floor in as few places as possible.

A reasonable method might be for a timber joist construction to be put together as the new inner floor and for that to be sat on shock absorbing blocks. They will dampen any structural noise and reduce any transference from outer to inner. Neoprene is a popular material which provides structural stability and vibration isolation. A block every 400mm to 1000mm in each direction will start to offer you the results you are looking for. You should consult with the neoprene manufacturer as to the best way to use their product.

Once the inner room is floated on timber joists, you should lay flooring on top. This will probably consist of 25mm of a reasonable density wood, like MDF or Plywood. Most likely this will be built up in 2 or 3 layers of 10 or 15mm deep boards.

As with walls the idea here is to create an air proof seal to reduce any sound entering via airborne vibration. You also need to try and have some mass in the construction to absorb structural vibrations.

Having built a raft of boards, to be your base foundation sat upon the neoprene pads, then build your joists, then place more boards on top, you will be left with a hollow construction that may resonate. This is easy to fix by stuffing it full of rock-wool or specialised acoustic wadding. This will reduce the cavity size, add mass to the construction and result in a better sounding room.

Lastly the construction of your inner room floor, must be solid enough to take the weight of the inner walls and ceilings, as well as all the studio kit. There will be some considerable weight here.

Don't forget that you will need to leave some space within this floor for running cables between studios.

Thick Walls

If you have ever visited a new house, built with cheep materials you will be familiar with how thin walls let sound pass straight through them. A person talking in one room can be clearly overheard in the next. If you have visited an older house with thick solid brick walls, you will know that very little sound makes it from one room to another.

The principle in action here is one of absorption. As sound pressure waves hit a wall they need to be absorbed to prevent them passing through the wall and getting to the other side. A low mass wall will vibrate and let sound pass through, a high mass wall will absorb the sound and prevent it from passing.

Again using the PopFM example, we are creating a timber room within a room. Starting with the outer room wall, a timber frame is a good structural support. Then apply a plaster board coating to both sides. Each layer of plaster board will offer you maybe 3 to 6dB of sound reduction. Apply 4 layers to one side of the wall and 4 layers to the other side and you will have 8 times that isolation. 8 sheets at a worse case of 3dB is 24dB of noise reduction or isolation. This is in basic mass terms alone. You could use a brick or concrete block construction and get similar results.

The second wall, or room within a room using similar construction techniques, will then offer you an additional 24dB, creating a total of 48dB which is very good.

This is working on pure acoustic absorption, and not on structural transference. Remember that the point of the isolation of the second room is to create a lack of physical matter for structural sounds to transfer through.

Your first outer wall is going to allow structural sound to pass through it. There are also techniques to help reduce this. If your construction team simply nail the 4 sheets of plaster board with one nail, through all 4 layers, the nail is going to act as a resonating pipe, allowing sound energy to pass directly through it. Whatever vibrations the outer sheet of plaster is suffering, that is going to be passed directly to the inner sheet.

To help avoid this happening to the hundreds of nails you will need to use. Consider different fixing methods that do not cross the entire structure. The first couple of sheets maybe nailed to the wooden frame, and then the last outer two maybe glued. This will significantly reduce any transference through the wall.

Again, as each sheet is applied, make sure that all corners are filled with either a plaster skim or a silicon sealant to close the gaps and prevent airborne vibrations passing through the construction.

Windows

Studios with huge great windows look fantastic and are delight to work in, however they offer some significant technical problems to your design.

- They reflect sound, just like a mirror reflecting light
- They are not effective sound isolation
- They are difficult to mount and very heavy to support

If you can build your studio without windows, you will have a much better technical design. Many people these days fit video links from control rooms to studios to provide the illusion of windows. It also means that one studio doesn't have to be next to another, or near the control room. Video links can make them all as remote as you like.

In reality a studio must have a window. Studios are a dead sounding space with a limited air supply, and taking away the windows may create a very claustrophobic space.

So the question now is how to have windows that don't leak sound, can be physically mounted, and don't simply act as mirrors, causing huge sound reflections and colouration of your studios sound

It is impossible to stop the windows causing reflections but they can be managed, so that the reflections caused are not a significant problem. The best way to do this is to angle the glass in the window so that it reflects any sound waves downwards. The idea here being that the floor of the studio is probably carpeted and that will absorb any reflection. It also means that as it another non parallel surface and you are reducing the chances of standing waves affecting the studio sound.

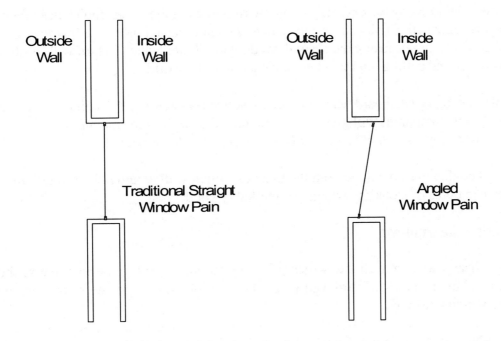

Figure 5 Window design

The diagram shows a single window, upon an inner and outer wall. The inner and outer walls offering isolation, however the single window breaches this isolation and will not offer any significant sound-proofing. The straight window shown on the left will also reflect any studio sounds back into the studio. The right hand image shows an angled window, that again offers little isolation, however due to its angle will reflect sound downwards toward the floor for absorption.

Double Glazing

A single sheet of 4mm glass will offer some isolation, but not very much. As air pressure waves (sound) hits one side of the glass, the energy of the sound is transferred into the structure of the glass. That causes the glass to vibrate in sympathy with the sound and in turn the air on the other side of the window will vibrate and the sound will be regenerated on other side of the glass.

A double glazed window is created by two sheets of glass being held together with a small air gap, normal sealed, and usually filled with an inert gas at low pressure. This low pressure gas filling is low in atoms compared to normal air. This means when the first window is vibrated by the external sound pressure the number of air molecules vibrating inside the sandwich is reduced, resulting in less sound transference. The low pressure gas then has a hard time causing the second window to vibrate, which in turn will result in almost no sound getting out the other side back into the world.

Double glazing is a standard way of reducing sound transference through a window, and good glazing companies will give you guaranteed figures for the amount of sound reduction their glass will achieve. 35dB is a common reduction level.

Fitting a double glazed pain to both the inner and outer wall, you will be heading for better than 60dB of isolation, which is remarkably good.

Mounting the windows

Glass is heavy, and the wall into which it is to be mounted needs to be able to take the weight. A hole in a plasterboard wall will not do the job. Make sure there are joists ready for the window to sit on.

As windows are transparent, the wooden support the window is to be mounted on may be in view after the job is complete. Make sure any final finishing is done, before inserting the pane.

To mount the window, place a final finish wooden baton along the surface that the glass is going to rest against. Remember as the glass is tilting it will exert more than the normal vertical pressure against the baton. Place rubber seals round the window edges so that the glass does not come into structural contact with the wooden frame. When the seals are in place, you can then place the final fixing batons to the window to clamp the glass in place. Ensure that all seals and surfaces also have a layer of silicon sealant at all points to ensure there can be no air gaps. Air gaps equal sound leakage. Fill them all, otherwise it does not matter how much you spend on high quality windows, the sound is going to simply leak through the edges.

Remember that we aim to have two separate walls with two separate windows to maintain the structural isolation. Make sure that any window sills do not join those two walls together. An illusion of a continuous window sill can be made by cutting the solid sill between the two windows and the space being filled with a strip of low density foam rubber, preferably in a matching colour to sill. Also make sure that any gaps at the sides and above the windows are masked with foam. This will also stop any dirt or debris falling into the window cavity which would look poor.

Figure 6 Double glazing

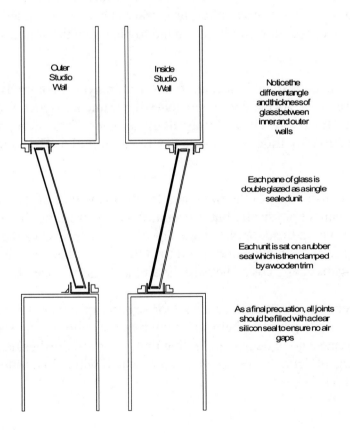

Outer
Studio
Wall

Inside
Studio
Wall

Notice the
different angle
and thickness of
glass between
inner and outer
walls

Each pane of glass is
double glazed as a single
sealed unit

Each unit is sat on a rubber
seal which is then clamped
by a wooden trim

As a final precuation, all joints
should be filled with a clear
silicon seal to ensure no air
gaps

Figure 7 Diagram of angled window design

Doors

Doors are key areas where lack of attention can result in very poor isolation and can easily cancel out all of your previous good work and intentions.

Using some lessons from earlier on you can do a good job. Air gaps leak sound. High mass absorbs more sound than low mass. Two doors will be better than one.

All you have to do now is put those lessons into practise.

Air Gaps

The door must have a perfect air tight fit to the door frame. Just as important is that the door frame itself is sealed against the wall. There is no point fitting an air tight door to a frame that itself doesn't seal to the wall. Whenever a surface to surface connection needs to me made, seal it, just in case. Then after the surfaces have been bonded, seal the edges again, just in case. Air gaps allow air borne sound to penetrate the studio shell, seal them for best results.

The main edges of the door should meet the frame with maybe a 10mm overlap. That 10mm will be a steel strip. As the door approaches the frame a magnetic rubber seal will clamp down on the 10mm steal strip, closing all the air gaps. This is very similar to the pliable rubber seals used by fridges to ensure that the door will not leak out all of the cool air.

Sometimes for the construction of the room and for the avoidance of doorsteps, the bottom of the door cannot physically butt-up against a steel or rubber lip. However, as the door closes, a lever within the door is activated as the door approaches the frame and can force a seal to drop from within the door itself to close the gap between the door and the floor. Close all air gaps at any cost. They are the single biggest cause of poor isolation.

As a worst case scenario, where professional door seals simply cannot be found, consider domestic draught excluding seals. There are many types of rubber strip or deformable plastics that will create a good door seal. Failing that, there are brushes designed to be added to the bottoms of doors to keep draughts out, which will help a little.

High Mass Doors

A thin door made of reconstituted wood and with a hollow centre is not going to stop a single sound. Indeed in the event of a fire this kind of door is more likely to speed the fire up rather than slow it down. Fire doors however are a very good choice of door. Although wood burns, a solid lump of wood will be heavy and massive. Better the sound isolation is achieved with thick and heavy doors. There are some heavy doors available, mostly fire doors, which can be cost effective. A true studio door is a specialist design and will simply cost a lot of money.

Studio doors often need to comply with local Health and Safety or fire regulations.

The ideal studio door will be at least 50 to 100mm thick. Based on a 50mm door this would consist of 10mm of wooden facia, both front and back, leaving a 30mm centre void, which should be filled with sand. The sand will provide the door with a great deal of mass, which will absorb sound and start to create the isolation.

Common problems with massive doors tend to be hinges. The door weight will always cause the door to drop and no longer fit in the door frame. Good practice is to fit at least four sets of hinges. Any less will result in severe drop, which will lead to poor sound isolation, not to mention significant cost having the door re-hung.

Once the door is hung, you should remember to oil the hinges with medium viscosity oil at least every six months. Any less will result in wear on the hinges again causing drop.

The door will also need to be fitted with an automatic closer, people never remember to close doors and if you have designed a sound-proof studio, you should make sure the door is shut otherwise your work will have been wasted.

If doors of considerable mass cannot be sourced, then modern UPVC doors and sliding windows are worth looking at. They come off the shelf, or can be made to order in various sizes, with very favourable isolation measurements of around 30dB.

The mass of the door will affect its resonant frequency. The lower the mass and density, the higher the resonant frequency. UPVC tends to be low mass, unlike a solid wood, or wood-sand composite. Lower resonant frequencies are achieved with higher mass. With luck below the audible sound ranges.

Two Doors

If you are employing two walls to create isolation, then of course you will need two
doors. Mass again is the key to both doors. Always remember to seal any holes in the
door frames. It may be worth having one door open inward and one open outward to
prevent clashes. If the doors are to be placed directly back to back they will need to open
in opposite directions.

Figure 8 Double door construction with viewing panel

Viewing Panel

Although windows add to leakage and reduce isolation, a door viewing panel should be
considered or even required by regulation. If you are expecting even moderate foot traffic
through the studio door, there is going to be a chance of two people trying to use the door
from opposite sides at the same time. There is going to be an accident. A viewing panel
(small window) allows you to see if someone is stood behind the door you are about to
open. Once a massive door starts moving, it is very hard to stop and an accident will
happen.

As with all windows in studios, they should be double glazed to reduce any leakage.
There should be a rubber or silicon seal round the window to ensure there is no leakage
round the window. The viewing panel will probably small, maybe 400mm by 200mm. As
the surface area is small, there is not much to be gained by tilting the glass in the door.

Air conditioning

When you create an air sealed room the oxygen in it is going to be depleted as people breath. The room is also going to get warm as people and equipment give off heat. That heat has to go somewhere. The temperature in the room is going to climb until either your equipment fails, or the flammable materials in your room catch fire.

The concept of air conditioning is straight forward and serves us two purposes

- Use some kind of refrigeration to remove heat energy from the air, and to keep the room cool
- Remove some old low oxygen air, and replace it with some fresh air for breathing

There are two types of air conditioning local cooling and forced air.

Local Cooling

The idea is to have an external cooling plant which supplies a coolant gas or liquid into the studio. In the studio a cassette unit, a set of fans and cooling fins, draws warm air over the super cooled fins, which reduces the air temperature, and results in the air being cooled.

This means that the necessary holes in the studio walls will be small the size of a couple of hoses, with one pipe for new fresh coolant and another to take away the used warm coolant. There is normally a third pipe for any water condensation that may form and need to be taken away.

This local cooling will result in studio noise as the cassette unit draws air across it and pumps coolant way. The "wind noise" can be considerable.

If the studio microphones are forgiving and very directional, you can place a very quiet cassette unit at a far end of the studio and you may get away with it. It will be much cheaper than forced air, and can therefore be an attractive choice.

You are still left with the problem of fresh air replenishment, which you must do. The details of which are covered in the forced air section.

The pipes that pass through the studio walls should be carefully managed. They should go through the smallest possible holes. Back fill them with plaster or silicon sealant to ensure no air gaps. There will be vibration noise passed down the pipes, so mount the pipes both inside and outside of the studio on supports which are structurally isolated or damped. This could be as simple as using rubber shock mounts to attach them to your building structure. Outside of the studio try and run the pipes away from any vibration or noisy parts of your structure, to reduce any vibration they may pick up.

When the pipes enter your studio, run them as little distance as possible. The longer the run in the studio the more vibration will radiate from them.

Forced Air Cooling

In most cases this is the best solution. It will also be complicated and very expensive. If it is too cheap, it will let you down, in both reliability and performance.

To create an air flow you must have a fan. The fins of the fan rotate and push wave after wave of air in a direction. The faster the fan spins the closer together those waves will be, until actually you have sound.

The way to avoid noisy air-conditioning is to have large fans running very slowly. The pressure waves they create will be at a frequency below that of normal human hearing, sub sonic. The secret to good studio air conditioning is large and slow.

The cooling in a forced air system is simple; house all of your cooling plant outside and away from the studio. Compressors create a coolant, which cools radiators and fins, in the air flow. A low speed high volume fan pushes the cooled air down a long duct into the studio. The studio being an air sealed box, must also have a return air flow, otherwise the pressure build up would simply stop the fresh air getting in the studio.

Fresh air must be added to the system to ensure used oxygen is replaced and this can be done in an air handling unit outside of the studio.

The job now is how to duct a large air volume into a sound-proof room without letting sound in, or out. The ducts themselves, one in and one out, will be large maybe 750mm x 300mm cross section for our standard PopFM example. To ensure as little structural noise as possible enters the studio; mount the ducting on shock absorbing pads and break the metal duct regularly, with large rubber joints. This will reduce any structural noise passing along the pipe. A key place for a rubber joint to be placed is between the inner and outer studio walls. Avoid coupling these two walls together at all costs.

Also consider airborne noise passing down the duct. This can be reduced by placing obstructive baffles in the airflow. The airflow, like water finding its way round rocks in a river, will continue flowing down the pipe and sound absorbing obstructions will reduce the airborne sound content.

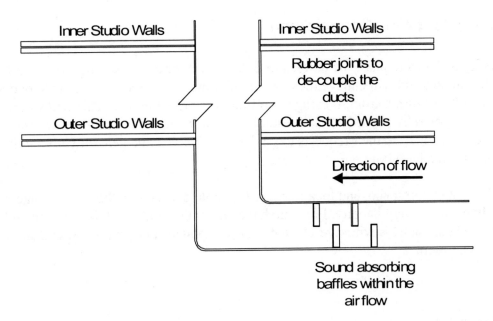

Figure 9 Isolated and damped air flow

The use of low speed airflow and baffles will be greatly enhanced by seeking advice from an air-conditioning or acoustic specialist.

The point in the studio where the forced air enters the studio is normally a ceiling grill. This should have directional fins so that you can point the airflow away from any critical equipment that may be prone to wind noise. Traditional grills will have many small fins and sharp edges creating eddies as the air passes over them, these eddies turn into noise. Larger grills will have a slower flow speed passing over the fins, reducing any wind noise.

The return air flow should similarly be acoustically damped, and structurally isolated, to avoid any sound transference into the studio. If you have several studios in a line, all of which go back to common air handling plant, you might want to think about sound transference in the parallel air ducts that could cause one studios sound to leak into another.

You may wish to provide some sound absorption to the outer case of the ducting to reduce any transference between air flows. Always go back and seal any gaps between the ducting and studio structure. Using plaster, where possible as it has mass, and where that's not possible, silicon seal or expanding foam.

User Controls for Air-Conditioning

It is never a good idea to place the temperature, fan speed, or on-off settings inside the studio. The studio users will believe they have a life time of experience on how to operate the system and will turn it from freezing to boiling every five minutes. This will shorten the life of the system by about 50% and result in regular complaints about the system not responding fast enough to their requests to go from 30C to 4C, and generally giving the engineering team a hard time.

Set it to 21C, slow fan speed, and lock it away. If people don't know they can change it, they will live with it, very happily. Let them have an opinion and it will be unreasonable and they will break it. This will save you thousands of pounds per year, in repairs and energy consumption.

Fire Detection

Most fire officers will insist that any room within a room is a risk. There are a limited number of exits and fire officers start to get nervous. You will almost certainly need to add an industrial fire alarm system into your radio station.

In each separate space, be it office, store cupboard or studio you should fit a smoke detector, wired back to a central control panel using fire rated (pyro) cable, This cable should be surface mounted to the plaster walls or ceiling of the studio and pass through the studio walls at a sharp 90 degrees, and the holes back filled with plaster or silicon. Close the gaps!

Having gone to great lengths to create quiet in your studio, many people fit flashing red strobes lights in their studios to indicate fire, rather than a traditional bell or siren. This will visually inform the studio staff of a fire, allowing them to quickly finish speaking on the radio, put on a tape or a CD and run away. A siren on-air would give away to the general public that you are on fire, and then you would become the story, which was probably not the plan.

Air-conditioning systems which force new or fresh air into a room should be linked to the fire alarm system. In the event of a fire the system must shut down the airflow and prevent any fires being fed by fresh air.

Make sure that you comply with local fire regulations.

Electrical Services & Cable Runs

Wherever the studios are, they will need to be connected to a Central Technical Area (CTA). Normally this would provide the studio with power, and a connection to the transmitter, as well as many other electronic services, such as sound and TV feeds.

It is important at the design stage to work out where the cables will run from the CTA to each studio. Will each studio have its own cable run or will there be a common cable route that all the studios follow? This is the kind of information that your builders will need, to make sure they put channels under the floor for your cables to fit in, or gaps in the walls for the cables to poke through.

You should know how much cable you will want to run to each studio, and this will be part of the electrical and electronic specification.

It is a good idea to make yourself a bundle of the cables that will go into each studio, maybe only a 300mm long, so that you can see the physical diameter of the cable bundle. This will tell you how much space you need to leave under floors, and how big the gaps in the wall need to be. It may be easier to lay the cables flat in a channel rather than bundling them all together.

At the studio end, work out where the cables will arrive.

If you have a rectangular room, and you are placing a mixing desk in the centre of the room, you probably want all of your cables to arrive under that desk. Possibly inside some kind of technical cupboard which is part of the woodwork construction.

If the cables arrive in the far corner of the studio and then you want to get them to your mixer in the centre of the room you will have to step over them 10 times a day.

Good planning at this stage will help you or the builder make sure they can get the cables to the right point in the studio by building in cable channels to the floor space.

You may also want to think about any future expansion, will it possible to get extra cables into the studio? Things change, technology changes and today's cables standards will be superseded by something new. You will need to get that new cable into the studio without getting the builders back to knock the walls down. Make sure the structure can accommodate new requirements.

Again, once the cables have been run, seal the holes. It is common to use small sand bags stuffed into the cable ducts. They can be moved when you need to add more cable in the future.

It may be necessary to provide more than one cable access route. Power cables and signal cables should be kept separated as much as possible. The power cables can induce noise into the signal cables, which will result in your studio humming or buzzing electrically, ruining any sound-proofing efforts you have made. You will want to keep your power and signal cables at the very least 500mm apart, ideally a metre or more. If power and signal cables need to cross over each other at any point, make sure they cross at 90 degrees. This will ensure as little electrical interference or cross talk as possible.

It is also worth trying to keep the cable runs straight, if you need to feed new cables into the studio space later the more bends your cable access has the harder that is going to be. It is also worth fitting a draw cord into the cable run so that you can use it to pull a bundle of cables through the route rather than having to try and push them along from a distance. Make sure the draw cord has lots spare at both ends of the route. Otherwise once you have pulled the cable through, you will never get your draw cord back to do a second job.

Within the studio itself, cables for electrical services such as lights should be surface mounted within conduit to the plaster walls, and not chased into the plaster. You need that mass for isolation, don't cut it away. The studio may have a wooden dado rail or a heavy skirting board behind which cables runs can be hidden.

Where possible try and use cable tray for any long runs. This is a metal or plastic tray that mounts to a surface, upon which you lay the cables. They can be tied down to the tray to keep them neat and tidy. Again cables and trays will transfer structural noise, so make sure the cable tray is broken across the studio isolation gaps, and where the tray will be passing through areas of high vibration, use shock mounting pads to reduce the pick up of vibration. Always make sure that the cable tray and any metal body structures are earthed for electrical safety.

Also remember before the cables are run, label them at each end. When you pull 20 identical cables through a hole or over a long run, trying to remember which is which is almost impossible, real names or numbers are just fine, but they must be uniquely identified. Try to use the same label scheme on the cables, as is used in the technical plan for the station, so that cables can be latterly matched to the design. This will help in fault finding.

Final Finish

Once your structure is in place you will need to produce the final finish or décor for the inside of your studio. Having taken care to ensure the room is the right shape; cabled well, air-conditioned, with good air tight doors, the final finish is what will give the studio its look and sound.

The inner studio plaster wall is hard, sharp and reflective. In the earlier section on sound colouration we talked about reducing reflections to reduce the RT60 reverb time. It is traditional to cover your inner plaster wall with a soft absorbent fabric. Build an additional inner timber frame standing 40mm proud of the wall. Fill the recess with a medium firm foam rubber, which is rated for sound absorption. Then cover that and the timber frame with a final finish fabric A tight knit Hessian can be stretched across the frame to create a solid wall, packed out with foam behind it.

There are specialist companies who will upholster you studio for you and make the place fit for habitation.

The floor is a great place to add an absorbent surface. It may look amazing, but a polished wooden floor is simply going to create reflections and fast reverberation. A carpeted floor will make the room sound much better. Office carpet tiles, designed to be anti-static will do a great job, and can be easily moved should access to the floor be required.

The ceiling again needs to be acoustically treated. Normal office ceiling tiles will work well, and offer a reasonable amount of acoustic damping, but there are specific acoustic tiles that are designed to perform much better, and increase the absorption.

Studio Colours

A waist level dado rail into which mains power, light switches and other cables can be hidden can also act as a break point for the final finish fabric. Large continuous sheets of fabric are very expensive, but smaller ones can be much more cost effective. It also means that you could have a different colour on the top half of the studio compared to the bottom.

Stations that have generic logos, such as just words or symbols might want a plain fabric finish across the whole studio. Stations who have a strong colour base for their logo, such as blue and red, might want to split the bottom half of their decoration into blue and the top half red. Also think about the station logo and how long you think the company will keep it. If it is printed on each wall, and after 6 months the station changes its name or brand, then you're going to have to pay to get that changed. Hanging a brand sign on the wall might be an easier way of dealing with this.

There are plenty of experts in what room colour and decorations are good for the human condition, For example, a blue room, will induce calm and precision, where as a red room will result in a fiery and passionate mood. A yellow room may induce a sense of nausea. A mixture may work well. Check out the local thinking. Colour choices made now will probably still be around in 5 years.

Wear and Tear

Look at the studio space and work out where people are actually going to be most often. They are probably not going to go in to the far corner very often. They are going to sit or stand at the mixing desk for hours on end, they are going to walk to the studio door millions of times, and they are going to lean on the wall next to the studio door, to help them pull the door open, because the door is heavy.

These will be your main areas of wear, and where the studio will start to look shabby first.

At the mixer, consider keeping spare carpet tiles which can be replaced every few years. Fit a hard wearing plastic or wooden floor just at that point. A large mat on the floor does just as good a job. By the doors, again look at the flooring, or adding push panels to the walls next to the door for people to lean on. People will lean on the wall, and if it's a soft fabric, padded with a foam, it not going to last very long before it stretches, tears or simply compresses and looks shabby.

Electrical Power

Specification for Electrical Systems

The basics for a system specification should be as simple as taking a building plan and drawing on where the power outlets should be. It is then up to the electrical consultant to create the rest of the specification according to its needs.

If you wish to be more involved then you can go all the way to drawing building plans with layouts and circuit paths, including breakdowns of power ratings for generators, UPS, cables, distribution boards and small power locations.

Remember that unless you are qualified in this field then you must pass this out to a contractor. Work with them to make sure you are getting what you need. When the project is completed they are required by law to provide you with safety certificates for the system.

Theory

Ever since the light bulb replaced the gas lamp we have consumed more energy year after year than ever before. In a radio station, power is everything; it's the blood in your system, without it, silence!

The size of your station will dictate how your electrical installation will take place. The choice at this early stage is single or three phase. You need to be able to work out your total power consumption to help you work out what kind of supply you are going to need.

As an instant rule of thumb, the following areas will requires supplies capable of:

- Studio Broadcast equipment 6amps
- Studio air-conditioning system 16amps
- Studio lighting 3amps
- CTA Rack of equipment 6amps
- Small office of 5 desks and computers 16amps
- Small office air-conditioning 16amps
- Small office lighting 6 amps
- Large office of 20 desks and computers 30 amps
- Large office air-conditioning 30 amps
- Large office lighting 16amps

These are only a very simplistic rule of thumb, and attention should be paid to the details of the system you are looking at, to ensure you understand your needs.

Single Phase or Three Phase

Power supplies from local electricity suppliers tend to come in standard ranges of 30amp, 60amp or 100amp.

This is the same for both single and three phase supplies, for example you can have a single phase of 100 amps, or a three phase supply of 3x 100 amps. This is a lot of power.

If your total power requirements are more than 30 amps, you should consider moving from a single phase to a three phase supply. The electricity supplier may insist upon it.

UPS and Generator Protection

For a station of some size where large amounts money are involved for playing commercials, or the stations credibility will be called into question for suffering power cuts, you may want to think about protecting your power supply systems. The local electricity supplier whilst wanting to deliver electricity to you all of the time, may not manage to do so. Local faults may cause you to have the odd power cut now and again. It is unusual if you are in a major urban area for a power cut to last more than a few hours.

To protect against such losses you might want to think about installing a Uninterruptible Power Supplies (UPS). UPS are essentially a very large box of batteries. They sit between your main building power supply and all of your broadcast equipment. If there is a power cut, for whatever reason, they simply continue providing power by discharging their batteries.

They come in a number of sizes from ones that will last just a few minutes to those that will last for several hours. The smaller units tend to be for IT equipment that need enough time in a power cut to shut down nicely without causing massive data corruptions. They might cost a few hundred pounds per machine. The larger units tend to run entire radio stations for maybe four hours, costing around ten to thirty thousand pounds.

A decent UPS may protect your station for as much as four hours, and few power cuts ever last that long. For complete protection you may also wish to add a generator. Thus in the event of long term power problem the generator will take over providing primary power to your systems.

If you do add a generator to your station, it should be one that starts automatically whenever power problems occur. This will also allow you to buy a lower cost UPS with a shorter battery life of maybe one rather than four.

Locations of UPS and Generators

The size of the installation and therefore the size of the UPS you have chosen, will play a significant part in where they are located.

A small UPS which is capable of supporting a small 2 studio radio station for only 2 hours might be only 10kVA. This is turn is quite small, maybe 1 cubic metre. This could be located in your CTA (central technical area) or racks room.

A medium sized UPS 30kVA, which could support maybe 4 studios and a CTA might be 3 cubic metres, and therefore might be too large to house in your CTA.

A large UPS 100kVA which could support 10 Studios and a CTA for 4 hours, could be as large as 8 cubic metres and probably will be housed away from your CTA as the size and heat generation is probably too much for the average CTA.

There will also be a weight issue, UPS are built around lead acid batteries and can weigh several tonnes. You will need to check the structural floor loading in the location you intend to place it.

It is therefore common for UPS to be housed in a basement or ground floor. It may also be prudent to check on the physical dimensions to make sure it will fit through the doors and corridors when it's delivered.

UPS are reasonably maintenance free and do not need continuous attention, so placing them away from the main part of the station is just fine. Check with the manufactures recommendations for locations and also check with local fire and health and safety regulations. If the UPS in a basement consider, ventilation, cooling and the chances of flooding. Water and Electricity do not mix!

Make sure there is emergency lighting in your UPS area, as when there is a fault working on it in the dark is miserable.

Generators are normally large and noisy beasts and will tend to live in garages, basements or outdoors in special housings. If the generator is outdoors you may need to get planning permission. There may also be a noise issue. You may be required to silence the generator if you are near any residential property, and that will start to cost money.

Ensure your generator is rated to at least the same rating as your UPS, and possibly higher. The nominal UPS rating of 100kVA will be its normal maximum power loading. If you are drawing full power and also recharging the batteries after an outage, then the UPS itself will consume more power. The generator needs to be able to provide this.

Also ensure that with the generator running at full load that the basic fuel tank will be sufficient for at least 24 hours. In the event of a serious problem the 24 hour tank will let regular fuel visits be organised. Again check regulations for where you can store fuel.

Storing hundreds of gallons of diesel may be a problem, which could force your hand to place your generator outside.

You also need to consider how easy it will be to re-fuel. Can a tanker simply drive up next to it and fill it up, or will there be difficult pipe work or hose routes?

On-Line or Off-Line UPS

An off-line UPS allows primary mains power to pass through it and run the systems that depend upon it. Should there be a problem the UPS switches the feed away and provides alternate power. When that happens there will be a brief power break of between a few milliseconds and a few seconds, as the power swaps from one system to another.

An on-line UPS is designed to ensure power is not interrupted even for a few milliseconds during a problem. The output is un-interrupted.

The power is rectified from AC to DC, and used to charge a bank of batteries. The batteries themselves form a DC power rail, which in turn powers an inverter to provide the AC mains power once more.

The advantage of the on-line system is that the DC battery power rail is always supported. If the mains drops the batteries continue to provide the power to the DC power rail, and the power output is maintained.

Wrap around and Bypass Systems

As with any electrical device or system, it must be considered that one day it may need replacing or repairing. The UPS itself offers a protection from power supply problems. There should be a method of protecting the supply from the UPS. Should it break then a bypass system should be used to work round the problem. This should be designed at the installation phase of the project, as once the system is on-line and in use, it will be too late to retro-fit.

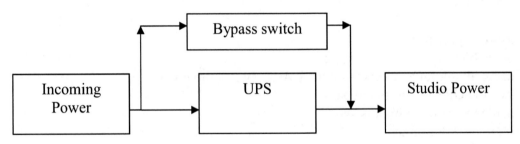

Figure 10 UPS bypass system

This allows the UPS to be bypassed for emergency work, such as replacing it, or doing basic servicing.

At the incoming power distribution board, have two breaker outputs, one to feed the UPS and one to feed the bypass switch.

This will allow power into the bypass system and also allow the input to the UPS to be isolated so it can be removed safely. The output of the bypass switch will be parallel with the feed to the studio power distribution. The UPS output should have an isolator and the feed from the bypass switch should have an isolator. This way the entire system can be bypassed safely and if necessary the UPS removed or repaired.

Bypass Procedures

Switching large amounts of power is a dangerous thing and can result at worst in explosions that could kill you.

UPS are designed with a number of useful features to help protect you. The output of a UPS should be in phase with the input. When you then bypass the UPS you will be essentially putting a parallel copper circuit round the UPS. If the phase of the input to the UPS is not the same as the output there will be a large bang, and you will lose all power.

To help with this most UPS are phase locked from input to output. Better than that, most come with an internal wrap around bypass. This will provide an internal soft bypass which will that guarantee phase in is locked to phase out. The hard external bypass can then be operated with be no chance of danger.

It is critical when the system is designed that you document the procedures to operate it safely. It may be many months or years before it is used and you will probably be under pressure when you have to. Therefore, make it easy for yourself with big clear diagrams and instructions.

Example procedure:

1. Ensure bypass switch is off
2. Enable breaker into bypass system
3. Ensure all output breakers from the bypass switch are enabled.
4. Place the UPS in internal bypass.
5. Enable the external bypass switch
6. Isolate the UPS input
7. Isolate the UPS output
8. Switch off the UPS.

With luck, the power is still running through the bypass to the studio. This is a complex procedure and an electrical consultant should make sure your procedures are correct.

To take the system out of bypass the procedure should be pretty much the reverse of above.

Bypass Safety Systems

Systems that switch power can be dangerous. So consider safety interlock systems. These are ones where in a procedure you must do point 1, which will release a key or interlock to allow point 2 to be completed.

There are also systems that will hold a key which will only be released when the UPS is in internal bypass, thus preventing out of phase supplies being connected together. The key is then required to enable the external bypass.

These will offer considerable improvements in safety in the system and should be seriously considered.

Again, get advice from a professional and check regulations and requirements.

Testing Bypass Systems

It is good practice to test all electrical systems before they are live and on line. When it next gets used it may be an emergency. You need to understand it and have confidence in your system and your ability to operate it.

However the first time the systems are tested get your electrical consultant to do it. They designed it, and have charged a fee for it. Get them to demonstrate the safety of the system.

Single Phase

One of deciding factors on single or three phase power is the size of the supply needed. A single phase supply of 100 amps may be enough to keep the station going, but you will find it hard and expensive to source electrical equipment rated at 100amps for just one phase. It is much easier to get electrical systems and components for a three phase supply rated at 30amps per phase.

Three Phase

A single phase power supply means there are working voltages of 220VAC, which is dangerous enough. Moving to three phase means that there are working voltages of 400VAC. This is the potential between the phases. As it is much higher more attention to safety during the design stage needs to be considered. Until very recently it was considered to be good practise to separate electrical phases into zones. One floor or large room powered by one phase, and the next large room or floor by another. Indeed in the Theatre and Lighting industry it has always been the practice that equipment running on different phases should be separated by two arms lengths. Spread you arms their fullest extent so your hands are as far from each other as possible, measure that distance, double it. The theory being that a 220 volt single phase shock is very dangerous and often lethal, a 400 volt three phase shock will kill first time. To reduce the risk, keep the phases apart by two arms lengths. The chance of being able to touch two phases at the same time is then reduced considerably.

Much equipment now works at a three phase level, and better standards of electrical safety and testing have been put in place to reduce the risks of shock. These are good things, and details of which can be found in the IEE 16th edition, as well as various parts of building regulations and Health and Safety guidance.

Who can do an electrical installation?

You will not be doing any electrical installation work, unless you are qualified. An Electrical Engineering degree does not qualify you. You need to be a qualified Electrician, insured and regularly inspected by local electrical bodies.

The best advice is, not to do any electrical installation work.

An installation is anything that is fixed device. So for example the lighting feeds, and sockets in the studios or office space are an electrical installation. They require qualified people to perform the installation.

Inspection and Testing

The regulations change frequently, and you need to check out the current situation, however, at the time of writing, an electrical installation should be inspected and tested at the time of installation. It will then be certified. Depending on the environment, there will be a requirement for a retest every 3 to 5 years. In most cases this will involve a power shutdown. Think about how you will keep the radio station on air during the power shutdown periods.

For portable equipment, anything that plugs in, there is Health and Safety requirement for inspection and testing. This can be performed by anyone who is "competent". A qualification in Portable Appliance Testing would make you a competent person. At the time of installation an item should be tested for safety, and also at regular periods of time depending on the environment of the equipment. For example, a photocopier in the corner of the office which never moves could be retested every five years. However a portable disco kit going out to gigs each night should be tested maybe once a week.

Visual inspections are very important, and should be performed at least twice as often as electrical testing. This is to give early warnings of problems. They may result in the need for electrical testing, or simply to initiate repairs.

There is considerable legal liability associated with electrical safety and it should be thoroughly understood or simply avoided and passed to professionals. Check out current regulations in your area.

Records

Records of all electrical work and testing should be kept for both practical and legal reasons. In the event of an accident a Health and Safety investigation team will demand to see records of all works done and of the regular testing. Without these records a presumption of guilt will be made. They are the first line of legal defence.

It is also useful to keep records of all works and testing so that you can understand how your systems work and make sure they are operating correctly. They will also be the starting point when adding in new equipment or repairing broken systems.

Over Specify

When designing any system that can be measured, has tolerances or limits built in, it is often prudent to over specify the needs at the design stage. If for example the design for a studio system will require 13amps to run it and you specify an electrical feed for that area of 13amps, there will be limited future expansion. If the studio suddenly needs a new amp or TV set and you don't have enough power, you will be simply unable to expand. If you have prudently supplied a feed rated to 16amps, you will have plenty of spare power available to continue growing and changing.

Earthing

All safe electrical systems should employ an electrical safety earth for two reasons.

- Acts as an electrical reference for signal measurement
- Provides a catastrophic electrical fail safe

An electrical voltage is often referred to a potential difference. The voltage measured is between two known factors. A battery has a positive and negative terminal and the voltage between them may be 9 volts. If only one end of the battery is measured, there is no difference as the other end is not connected. A bit like distance can only be measured between two physical points.

In electrical systems there must be two points to measure a voltage. If a CD player is connected to an amplifier, a signal which represents the sound from the CD and also a reference (earth), must be provided for the amplifier to work. Without both of those two points your system will not function.

For larger systems, where a signal travels large distances, for instance from the studio via a technical area to a transmitter some floors away, a common reference must be found for the two devices to talk usefully to each other. That is the earth.

Whilst more complicated, electrical power systems, work on the same basic principle. There must be two points for power to flow between, and normally this would be the live point via some equipment to the neutral point. The neutral point is separate, but usually the same (electrical potential) as the earth point. Those two points allow the electrical system to work.

In the event of an electrical fault people and equipment can become damaged, or cause a fire. Any of these things is a disaster, so all electrical systems should be designed to fail safely. There are a number of ways to do this. Fuses or breakers can detect when an unusual amount of power is being used and fail in a way that shuts down the electrical systems. Residual Current Device's (RCD) can be used to detect when power is missing from an electrical system, when it leaks out of the circuits into people or earth. It then shuts down the power to try to limit any damage or risk to human life. In the worst faults where a live circuit comes into contact with a metal surface an entire room or building can become electrified. The idea of earthing is that all metal parts of the system or building are earthed. Electricity flows from a higher potential to a lower one, and hopefully not through you! If an electrical conductor does come into contact with any kind of metal like the metal case of an amplifier it would make it lethal to touch. If the case is earthed a large current will flow from the conducted through the case and down to earth. This will cause a fuse or breaker to detect the unusually high current and fail to a safe position. This is known as a catastrophic fail safe, the fault is isolated and rendered safe.

Clean Technical Earth

Many electronic systems rely on passing small signals from one part to another. As previously stated these signals must in most cases be electrically referenced to something, and often earth is that reference.

Earth is assumed to have a potential of zero volts. If your system is sending a signal of either one volt or zero volts, then to measure the received signal you compare the signal to earth, which should be zero.

Signal Sent – Earth = Signal Received

If the signal is one and the earth is zero and, then 1-0 = 1. The received signal is one. The system has worked. If you have a "dirty earth" and its potential is incorrectly 1 volt compared to the real earth of zero then 1-1 = 0. The received signal is 0, which is incorrect.

This example is very common in complex electrical systems and needs to be avoided. A common example of this is when you hear an audio signal that has "mains hum" on it, a low 50Hz buzz that has been introduced into the system, usually by dirty earths.

Dirty Earth

Any number of factors can contribute toward the earth becoming dirty. One common reason is large industrial plant, such as lift motors. In order to turn a motor capable of lifting heavy items such as a lift, huge amounts of power are used. Electric motors generate magnetic fields to turn fixed magnets. These huge magnetic fields need to act over a few inches inside the motor. They will still have a significant magnetic affect several metres away. If you have any metal item which is earthed in the location of such a motor, the magnetic fields will be passing through that item.

A magnetic field passing through a conductor will induce a voltage into it. This is the basis of electromagnetic theory and how generators work. In an environment where you are not intending to generate electricity it is therefore a problem. Any earthed metal nearby such industrial plant will conduct that induced electricity down to ground.

Modern buildings are large and complex, and the distance the earth cable must run from any one point down to the genuine original earth point, may be considerable. The genuine earth is the one provided by the electricity provider or an earth spike.

Cables and conductors are not perfect and resist the flow of electricity to some extent. Whilst the earth will be conducting the induced electrical noise from the point of its generation down to earth, it will encounter a resistance.

$V = IR$

Ohms law states that Voltage (V) = Current Flowing (I) multiplied by Resistance (R).

If the long earth cable has one ohm of resistance, and the induced power into the earth cable is 1 amp, then the voltage on the earth cable at the point of the induced noise will be 1 volt.

In the previous example, we are trying to measure a signal of 1 volt between two systems and we calculate the received signal to be:

Signal Sent – Earth = Signal Received

Then the sent signal of 1 volt, minus the earth which is now 1 volt = 0 signal received.

It is common practice therefore to run two separate earth systems in complex electrical and electronic systems, a clean and a dirty earth.

The dirty earth, which is the one which will become noisy, is often referred to as the safety earth. It will act to protect life and systems in the event of a catastrophic failure.

The clean technical earth is a separate earth system, carefully designed to be noise free against which signals will be referenced and measured.

Implementing dual earth systems (clean & dirty)

You need to identify the best noise free earth you have available to you. This would normally be the incoming point of the electricity providers feed. It could also be a genuine earth spike in the ground.

Seek advice

Get the advice of a competent electrical consultant as it will have both technical, operational and legal implications. Get this wrong and you may be earthing the entire national power grid on behalf of your electrical company and they may not like that, or you may be providing an earth that is not actually referenced to ground and render the entire system useless, as well as having removed a catastrophic safety earth. If there is an accident and someone gets hurt, you will be legally liable and rightly so.

Basic principle of technical earth

Find your best safety earth point, as far back in the system as you can, probably the incoming earth point of the electricity provider. At this point create a junction and insert a large copper earthing bar. This will act as the technical earth *star* point. Run a large gauge cable to each technical area from this point.

If there are 2 studios and a Central Technical Area, then you should run 3 separate earth cables back to the star point and join them to the best earth. This will provide a quality earth in all 3 of these new locations. Assuming the cable is not running hundreds of metres then a 16 or 32mm cross sectional area earth bonding cable should do the trick. The larger the cable the lower the electrical resistance and the better the earth will be.

In each of these new locations such as the CTA create a new local star point. From this point each local item that requires an earth will be fed back to this local star point. Every item in your technical system should now have a good quality individually provided earth cable. In a local area such as a racks room, you can now use smaller gauge earth cable of around 3mm cross section. Check with the electrical consultant for best practice according to the device being connected to.

Earth Loops

A large earth cable will have a very low resistance and will pass energy to ground easily, avoiding the problems of noisy earths as previously discussed. An earth system where loops are created will start to introduce problems.

Creating a loop within the earth system will result in that loop acting as a transformer coil. This will cause induction from any electrically or magnetically noisy system, such as mains feeder cables or lift motors.

This earth loop will then start to have excessive spurious noise on it, rendering the zero volt earth reference dirty. This could result in mains hum or even pick up of an AM radio station!

Remember as you add a signal onto your earth reference it will modify the main signal itself. In the worst case you can end up broadcasting someone else's radio station, because your electrical system is acting as a giant radio receiver.

How to make an Earth Loop

Make a star point earth as previously discussed. Run two earth cables to two different technical areas, such as a studio and a CTA. In the studio have one CD player. In the CTA have one amplifier and a speaker. Make sure that both bits of kit are earthed using you star point technical earth. Take an earth referenced connection cable from the CD player through the wall into the CTA and plug it into the amplifier.

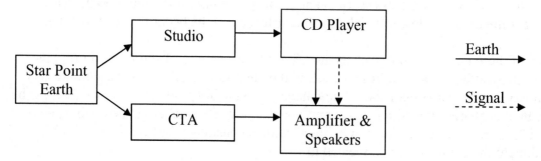

Figure 11 Generating earth loops

The signal cable will consist of the signal and an earth reference. By connecting the two clean earths together you have generated an earth loop.

How to avoid Earth Loops

Remember that the clean technical earth is a reference across the entire technical domain. So there is no need to pass the reference earth from the studio to the CTA, you can simply pass the signal only. It will still be referenced to earth when it gets to the amplifier because you have a clean technical earth in your site.

Often people will want to screen or shield a cable passing from one area to another, to prevent interference. The obvious problem here is that this will mean you are passing an earth from one location to another, and potentially introducing a loop. To avoid this simply disconnect the earth or screen at one end.

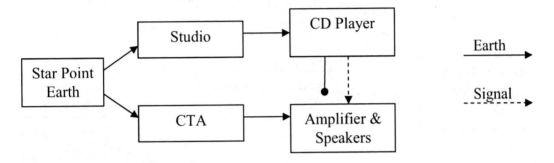

Figure 12 Avoiding earth loops

All inter area cables that have a screen or earth built into them should only be earthed at one end.

Convention suggests that the end that is generating the signal is the end that has the earth connected.

This means that the screen of the cable or in a very few cases the safety earth, is connected and functioning. The system works. By only having a connection at one end, you are not creating an earth loop.

Care and attention needs to be paid to electrical power systems where earth's are being disconnected at one end, whilst preventing loops, it is critical that a clean earth, of sufficient specification is connected to the new point, and that the equipment is clearly labelled for clarity and safety. Again seek advice from a consultant.

Some systems and devices are not referenced against earth but do require screened signal cables between individual items. As these systems are not earth referenced the screen should be connected at both ends of any cable runs as suggested by the installation manuals.

Complex Multiple Location Earthing

Where the electrical system is becoming complex you need to be clear about the design and intentions.

The star-point principle is the key and as the system expands the star point principle will need to expand with it.

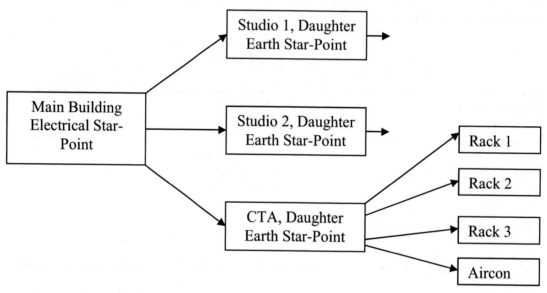

Figure 13 Star-point earth system

It may be useful to add a local star-point in each area. In the CTA for example, you may wish to run a clean earth from each rack back to the CTA star-point. The CTA star-point in turn is fed from the main building star-point.

Multiple Supply Clean Earth

When your electrical system becomes complex, then additional care needs to be taken to ensure safety. It is possible to provide 2 electrical systems to one area for redundancy, where one would take over the electrical load should the other fail. Care needs to be taken to avoid earth loops.

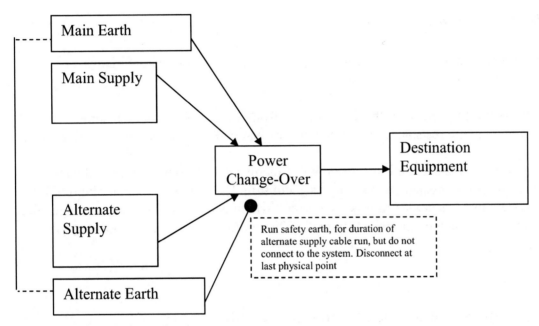

Figure 14 Dual supply earth system

The two separate electrical systems should have a common earth, bonded at the star-point. Along the cable runs themselves the safety earths should offer the same amount of protection. At the point where the two electrical systems meet the Power Chang-Over Unit only one earth should be connected, offering safety earth to the unit. The second earth should not be connected as this would create an earth loop.

A loop would induce noise into the system as previously described. The safety is needed on both cables (in case one is cut or damaged), and should both be connected at the star point, but only one provided to the equipment. This will offer a clean earth, without loops, whilst also offering safety earthing to the system.

Consultation

Systems such as these are complex and are often not catered for in the normal electrical safety regulations. Contractors will be un-easy disconnecting earth systems. To ensure the design is correct and safe, get an electrical consultant to check or design the work and have the consultant direct the contractors. You have a legal and safety obligation to get this right. The argument you are making is a technical one which is "superior to regulations" and you must be correct, otherwise in the event of an accident and investigation you will considered liable.

Earth Impedance

Any electrical circuit has resistance and impedance as part of its characteristics. A technical earth system is no different.

By using a very thick clean earth cable running from the star point to the technical areas you are reducing the resistance and impedance of that cable and making the earth cleaner.

The higher the resistance the less current will flow in a catastrophic failure, reducing the chance of fuses and breakers making that system safe. It also means that any item of kit on the system that has any earth leakage current, which is normally low, but manageable, will be putting a voltage onto the clean earth. That in turn will turn up as noise at some inconvenient point.

Only Technical Kit on Clean Technical Earth

Once a good clean technical earth has been created, protect it from any potential noise sources. So do not allow items with significant inductive loads to sit on the clean technical power systems. Anything with a large motor in it, like fan heaters, lift motors or pumps should be kept away.

It is very common for studios and technical areas to have 2 power systems delivered to them, a dirty mains system for dimmable lighting, heating etc, and a clean power system for studio equipment.

It is also common in studio areas, where film crews and cleaning staff will plug into any spare sockets to make their lights or vacuum cleaners work, to have two types of sockets. Use standard sockets for dirty mains, which anyone can use. This will be ideal for cleaners, phone chargers, and film crews. Use a different, non standard socket, for clean technical power that prevents people from accidentally plugging into your clean protected power systems.

Power Distribution

Once the power requirements for the station have been worked out with plenty of spare capacity for future expansion, backup and bypass facilities, the power needs to be distributed to the different parts of the radio station.

Normally the main incoming power feeds will be to a basement or some inconvenient part of the building. Basements are often good places for UPS but the local power for the studios and CTA needs to be more local, so it can be worked on and you can see the results of your actions in the same convenient part of the building. There is nothing worse than switching a circuit off in one part of the building, expecting only a few lights to turn off, to return a few minutes later to find an entire studio has been shut down.

Take the main feeds either single or three phase along with the star point earth, from the UPS, and run it to the CTA. The feeds terminate in a wall mounting Distribution Board with a main incoming breaker for the board and then smaller individual breakers for each of the circuits that will feed off from it.

If three phase power is being used, work out how to balance and distribute your power. The ideal scenario is where each phase is drawing a similar amount of current. Good forward planning will help. A standard power split for the three phases might be;

1. Studios
2. CTA
3. Heat, light, air-conditioning

Industrial plant within your building such as lighting, lift motors or heating systems will draw large amounts of power from and induce significant amounts of noise into the electrical systems. It is best to try and keep all industrial plant on a separate electrical phase from the studio systems and CTA equipment.

One rack of kit will probably pull about the same power as a small studio. So it may work out that; Phase 1 powers three studios, Phase two powers 3 racks, and Phase 3 powers lights and air-conditioning.

Distribution Boards

Distribution Boards come in a wide variety of shapes and sizes, an average sized one may be 0.4m by 0.7mm. When planning where to place them in the technical areas, and make sure there is enough wall space to mount them. They are heavy, and once all of the power cabling it attached will be very heavy. The power cables that feed them are usually thick (maybe 30mm diameter) and will come with a heavy steel armouring. Make sure the wall where the power is mounted, is structural and not a partition. Partition walls may well collapse under the weight unless they have been re-enforced. Ensure plenty of space around the board to bring in the cabling. Normally the cable will enter the board from above or below, on cable tray fixed to the wall. If the cables are presented from the side, leave plenty of space for the thick armoured cable to bend through 90 degrees to get access from above or below.

If you are using three phase, then there is a choice of running a single three phase board, or alternatively run three single phase boards, each board running only the one phase. There will be safety implications doing this, and advice should be sought.

Figure 15 Mains distribution panel

These boards are only handling the power. They will have safety earths connected to them. You should ensure that you are using a separate clean star-point for your system earthing.

Often the main power feed cables will be armoured and earthed at the source end. This will provide the safety earth to the distribution boards themselves. The board should take a new clean safety earth from the star point, and present this to an isolated earth bar within the board. Make sure that any feeder cables from these boards carry a clean star-point earth and not the dirty safety earth. Avoid connecting the clean earth to the dirty safety earth in the board as this will create an earth loop. Again get this right and seek professional advice.

Choosing Breakers

It is important to choose the right breaker for each board and circuit spurred off it. The aim is to ensure that the system is over rated to allow for expansion, whilst providing safety cut offs for any catastrophic failure. If a studio circuit will draw 10amps, then a 15amp breaker will give some overhead and space to expand and offer a realistic safety margin. Fitting a 100amp breaker would give you plenty of expansion, but in the event of a failure would probably allow the station to burn down before the power is shut off. The maximum breaker size should be lower than the cable rating it is feeding. That way any fault is protected at the same level as the cable is rated.

Breakers are designed to stop the flow of current and isolate a circuit when a set maximum has been achieved. A 15amp breaker will continue to work just fine at 14 amps, but will shut down at anything over 15amps. Designs and specifications change regularly so you check out the latest specs before ordering. Typically a sudden surge in power consumption on a 15amp breaker circuit, taking the load from 10amps to 20amps, will cause the breaker to shut down and the circuit to become safe. An increase over several minutes from 10amps to 18amps, may not result in the breaker operating at exactly 15 amps, it may not operate until 18 or 20. This is defined by the speed of the breaker.

Circuit breakers are available in fast, medium and slow speeds. The main breakers to the whole board which may be rated at 30, 60 or 100amps, should probably be slow. Smaller circuit breakers should operate first in the event of a problem and isolate the small local problem, and not take off the entire power distribution. Therefore the smaller local circuit breakers should be medium or fast speed.

Fast breakers can cause problems with things like lighting circuits where as a bulb fails it may momentarily short circuit. This is perfectly normal and will result in the element completely burning out and making it safe. A fast breaker will blow, and protect the circuit, but also will plunge the entire area in darkness as all the lights go out. There is a balance to be made and a medium speed breaker should be used.

RCD trips can also be employed, which monitor any difference between the live and neutral currents, if there is a difference it will isolate the circuit and make it safe. Principally these are a good idea, however if there is a lot of electronic equipment, there is a reasonable chance the cumulative "earth leakage" of the all that equipment will cause your breakers to fail safe, when actually the system is perfectly normal. Careful consideration should be given to industrial applications where good electrical safety is part of the design. They will increase the technical risk of power loss, but may increase personal safety. Seek advice before committing.

Dirty Power

As with the clean technical power systems, care should be given to the design and installation of the dirty power systems. Dirty power is any other mains power system that is not the clean technical system.

All the same standards should be applied to dirty power, in terms of breakers and designs. Consider using RCD trips on the breakers for added safety to non critical broadcast systems.

Avoid joining the two systems earths together any-where other than the main building star point. Be careful with signalling cables from dirty systems to clean systems to ensure that no earthing is ever passed from one system to the other.

Label Everything

No matter how good the documentation, make sure that during the install, every single socket, switch and distribution point has a clear and simple label that identifies it and where it is fed from.

An example label might read something like:

Pfm-std2-ls1 DB-UPS-01

Which means, PopFM, Studio 2, Light Switch 1 and is fed from Distribution Board, UPS number 1

This means you can look at this light switch and identify which Distribution Board feeds it, so it can be isolated and made safe to work on. As systems grow past a few simple interconnecting parts it becomes almost impossible to hold all of the information in your head and clear documentation is critical.

Dual Mains Systems

For many different reasons power systems shut down. It may be a fault, or it may be by design because of the need to work on the system to upgrade it. It may be the regular inspection and testing cycle that is in place, either way the power is going to go off.

Decide how critical the power is. If the station is a 24/7 operation then power needs to be available 24/7. Reality says it will go off, and the design should accommodate this.

Inevitably there will be a failure of any system and we have already looked at methods to protect against many of the common problems, using UPS and generators.

One additional option is a dual power system. This is where the main power systems are duplicated, nominate one of them as a primary and the other as a secondary.

Each item of kit takes two power sources. In the event that one fails then the second will take over. This has a number of advantages

- Back up power in the event of a fault on one system
- A secondary system to allow complete shut down of the primary for expansion or repair
- Increased power availability, critical in a 24/7 operation

There are of course also disadvantages

- Double the cost at time of installation
- Increased complexity
- Double the maintenance costs
- Potential for earth loop problems

As with all investment it is necessary to weigh up the risk versus cost. If the station is large and earns several thousand pounds per hour or per day in commercial advertising, then that needs to be protected. There may also be a brand image to protect at all costs.

If you can calculate the loss of income or credibility through a power failure you can work out if it compares against the cost of running two power systems. In most cases the comparatively small cost of installation can be earned back in the few hours you may have gone off air due to a single system fault.

Ideally every single power source and destination would be duplicated for total protection. However the reality is that cost will be an issue and it should be possible to prioritise certain areas. For example the CTA and main studio should have dual power, and maybe the back up studio can live with only a single power system. The risks of two faults occurring at the same time, such as a studio problem forcing the use of the backup and a power failure are low.

The cost of buying equipment with dual power supplies can also be considerable, but again there is a half way house.

Most studios and CTA racks use a device called a Mains Distribution Unit (MDU). You provide one power feed to the top of the rack, and it provides several fused power outputs. Then each item in the rack or the studio plugs into this distribution.

Dual Input MDU

Figure 16 Rack based mains distribution

MDUs come with two sets of power inputs, a main and reserve. You provide your best most protected and smoothed power circuits to the main input and a secondary supply to the reserve side. Normally all power is taken from the primary side, unless there is a fault when it changes to the reserve feed. That fault could be caused by intentionally shutting down the primary side for maintenance or upgrades. The power swaps over to the reserve side and the equipment remains on air. The beauty of this system is that it means you can use equipment that has only one power inlet, as most kit does, and feed it with a dual protected supply.

Intelligent Mains Distribution

Modern MDU's, dual powered, or not, can also be intelligent and be controlled by a WAN connection from a web page, or by an RS232 feed. This means that you can remotely monitor the state of your power systems, or receive an email or SMS text message in the event of a problem. Additionally if you are facing a sustained power cut and the UPS detects some problems then the UPS and MDU's can talk to each other, and based on configuration information you have previously supplied, start shutting systems down to save power. For example in an extreme situation, where the UPS is running off batteries due to a power cut, and there is no generator, it maybe decide that the main studio only needs two CD players and not four. Two of them can be shut down automatically to save power and start extending the life of the station whilst running on UPS batteries.

Mains Isolation Transformers

Years ago before the safety features of RCD devices were invented, it was necessary to provide additional safety to engineers who needed to work on live electrical equipment. For example, designing or fixing a power supply on a bit of kit. One method of adding safety to the workplace was to isolate the mains supply via a transformer.

All electrical potential requires a reference point. A battery must have a plus and a minus; a mains power system has a live and a neutral. This enables current to flow from one point to another. If there is only one point then no current can flow.

Mains power systems normally have earth and neutral linked at some point so that they are both at the same potential.

If you are working on a live bit of kit such as a power supply and you touch the live circuit, it is referenced to earth, which is the same potential as the ground you are standing on. The power finds you at the same potential and flows through you. You get an electric shock.

To help protect people from this, it was reasonably common to use an isolation transformer to remove the link between earth and neutral and to provide an unreferenced mains supply. If you touch a live wire, it is no longer seeking a current path back to earth, either via the metal case of the box, or the ground you are standing on. This was a good system and saved many lives. Indeed if you were brave you could hold a live connector in your hand and not feel a thing.

In today's world engineers may find more solace in using RCD devices to protect their work areas. They are designed to detect power loss from a system (over 30mA), which could be escaping through you. After 30ms the system shuts down entirely.

The world of isolation transformers did have one flaw which was if you touched the live and the neutral then you would receive a shock and no amount of earthing or RCD trips will ever protect you.

Isolation transformers are also very expensive. They are still commonly used on building sites and one of the reasons for this is that RCD trips are very sensitive, and do not work well with large inductive loads such as drills or motors. As such isolation transformers are the next best safety choice, but generally in small power environments RCD trips offer excellent protection at a very low cost. Again talk to the electrical consultant about the options for your engineering workbench safety.

Broadcast Systems

The size of the station and its purpose will dictate the type of broadcast equipment chosen to make the station a success.

Domestic versus Professional

Items of equipment such as CD players or headphones will have a huge cost difference between a professional version and a domestic version. A domestic CD player can be purchased for around £100, whereas the professional version may be 10 times that cost. Decide which type kit is right for you.

There are a number of advantages to professional equipment:

- robust for constant aggressive use
- run 24 hours a day, every day
- can be maintained and repaired
- have multiple output formats to interface with a range of other equipment
- can be operated remotely using wired controls rather than simply an infra-red remote
- fit into a cabinet or rack mounting system of some kind

Domestic equipment is designed for one task, with gentle usage and normally is not very flexible. If you are building HiRize FM then maybe due to the budget restrictions and the low use, domestic equipment could work well. Building anything larger such as PopFM or NewsTalk Radio then professional is the only choice, and the budget must reflect this.

Domestic kit can be a good choice in a very small station. The repair costs of professional equipment can often be more than the total replacement cost for a domestic item.

Balanced versus Un-Balanced

Most domestic equipment is un-balanced, and most professional equipment is balanced. Balanced audio is a system where the transmission of a signal from one area to another is achieved by sending the signal twice. The first signal is sent as per normal and the second signal is sent out of phase, similar to a mirror image. This is often known as Phase & Anti-Phase, or Hot & Cold.

Balanced has two main advantages.

- The signal is very well protected from electrical noise between the source and the destination. A simple electronic calculation at the receiving end results in a good quality signal being recovered from all but the worst environments. Noise induced into the cable system from the environment will be induced into both the phase and anti-phase signals, the recovery calculation will then subtract noise from both signals and ultimately cancel it out.
- The signal does not need to be referenced against a common earth, as the phase and the anti-phase are referenced against each other. This means that earth loops can be avoided in large installations.

Where ever possible choose balanced audio equipment.
Balanced signals work in both the analogue and digital audio domain.

Analogue versus Digital

Signal to noise ratios (SNR), the ratio of difference between the intended signal and the background system noise, is typically noticeable as the background hiss that can be heard on an audio system. When many systems are chained together, such as a CD player into a mixer, which in turn plugs into an amplifier, the result is that the noise is increased at every stage. The noise of each stage is reproduced and added into the next stage, which in turn adds it own noise. Very quickly a poor quality system is audibly poor.

Most modern top-end broadcast equipment is digital, and the audio quality results are spectacular. SNR in digital systems can be measured in the region of 90dB or above, which is excellent. In analogue systems the SNR of balanced audio equipment may be between 60 and 80dB. In unbalanced analogue audio SNR is at best around 60dB and that depends on the use of the equipment and how far the cables run between bits of the system.

Digital systems start with incredibly low noise and can, depending on the system, still result in cumulative additions of noise, but as they are all very low, will result in a final low noise system. Digital systems are generally considered not to introduce any additional noise.

There are pros and cons to both digital and analogue systems:

Digital Pros
- Low noise
- Exceptionally immune to noise
- Reduced cable installation, one digital circuit will pass both L&R audio channels
- Often very flexible in terms of configuration
- When standards and levels have been agreed, rarely need to be checked or reset

Digital Cons
- In complex systems can be hard to make work, due to clocking, jitter and synchronisation issues
- Very difficult to test without expensive test equipment and some considerable knowledge
- Very hard to test in service, testing tends to stop the signal which is catastrophic if the equipment is on air at the time
- More expensive
- Reduced variety of digital equipment, although this is changing

Analogue Pros
- Lower cost
- Easy to test in service, analogue circuits can be listened to in parallel using simple test tools such as headphones
- Easy to understand, and potentially possible to fix

Analogue Cons
- Signal to noise not as good as digital
- Offers less power and flexibility in complex systems
- Requires regular line up between systems for compatibility

Digital systems offer the opportunity to simply reconfigure or remotely control operations. If you have a problem at night, then it could be resolved via a laptop from home. Analogue does not offer this option.

Digital configuration is complex and requires detailed understanding, and requires some considerable experience, not the domain of a novice, but given time and understanding the power is amazing.

With the right choice of equipment, the digital world offers flexibility and control at a reasonable price. The ability to expand into the future is where you want to be.

Planning the Specifications

As with most parts of the design and build of the radio station it is critical to get the specification right. You may be contracting some or all of the work out, or you may simply need a detailed technical plan for yourself to build the broadcast systems. Either way, a clear plan with plenty of detail is crucial.

There are many ways to build a radio stations broadcast systems, and many different systems to choose from along the way. You need to have made some of those choices by now, especially if you are about to start building the station yourself.

To write a specification, start with a large blank sheet of paper and draw a simple block diagram of how the basic parts of the station will work. Then take each of the block sections and break them down in more and more detail until eventually you have drawn every item of equipment and every connecting wire.

Here is a basic block diagram of the PopFM example:

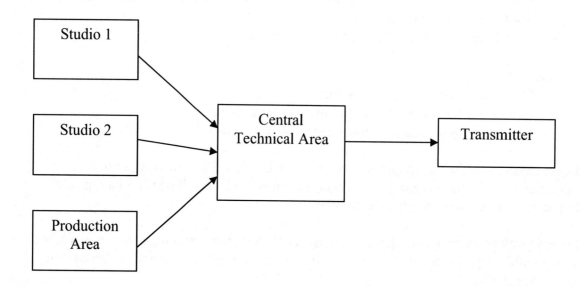

Figure 17 Basic block diagram of studio transmission chain

The diagram shows that there are two studios, a production area, all connecting via a CTA which then links the station to a transmitter.

The next stages should break down what happens in each area, our equipment list for studio 1 in PopFM is as follows, and introduces many broadcast techniques and systems.

Studio Specification

The studio specification will help you understand the detail of how the radio station is going to operate. Start by listing the equipment you will need to use, and then by connecting it all together.

As the system is drawn out, questions will be raised about technical standards, methods and designs that we will address as we go.

Professional 8 to 12 channel audio mixer
2 Broadcast quality microphones
2 Professional CD players
1 Minidisk player/recorder
1 DAT machine
1 Amplifier
1 Set of speakers
2 Pairs of headphones
Distribution amplifiers
Outside source selectors
TV
Email and internet PC
Interconnecting cables
1 Computer play-out terminal
1 Telephone phone in system and TBU interface unit
Jack Field
Mains and power distribution and management

We can now draw out a system diagram for the studio. Some items do not appear on this diagram such as the internet/email PC as it is a stand-alone device, which is not connected to the mixing desk. It does need to be included in the build, but not in the audio part of the specification.

As we look at each area we will see the functions the mixer needs to have. There are many mixing desks available and only a few of them are designed for radio broadcast, the rest are for nightclubs, concerts and music studio recordings. Whilst similar, the broadcast specific features are the bits that are going make your life easier and the presenters happy.

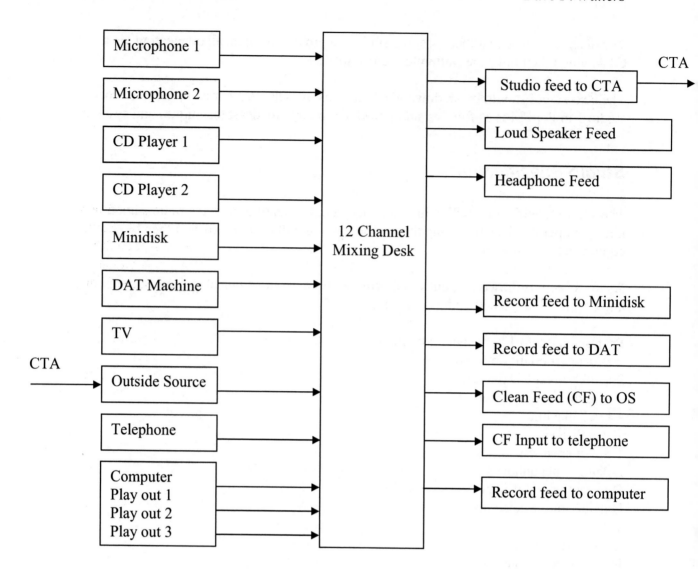

Figure 18 Studio mixer and outboard equipment

This is the first time we have looked at a detailed system diagram for a studio setup and it has immediate thrown up some new information, terminology, and types of kit, that will now be addressed.

Line Level

Line level analogue audio is measured so the signal with a pure sine wave will peak and trough at plus and minus 0.7 volts around a central zero volt reference. A line level device or input is designed to output or accept a signal where normal operation is based on this measurement. This is known as 0dB.

Most audio systems are designed to work at this level as well as provide some headroom. This is where the normal level of input at 0dB may occasionally peak above this. Audio is dynamic and sudden bursts of extra volume have to be accounted for. A sudden drum, bang or cough could typically be two or three times louder than the norm. Most audio systems in the analogue domain will allow for an extra 22dB above the normal 0dB for such things.

A mixers line level input will normally have a gain control which will allow correction of a signal which is not providing a 0dB level. This is usually around 12dB either up or down and is a control on the mixer channel. This can then boost quiet signals and reduce loud signals.

In our example the mixer is expecting a number of line level inputs, such as CD and DAT machines. These will be line level inputs.

Microphone Level

The mixer must accept microphone inputs. There are two main types of audio signal microphone level and line level. Microphones provide a signal output which is very low, typically 40 to 60dB lower than a normal 0dB line level.

The mixer must be able to provide the extra 60dB of gain to make the signal from the microphone comparable to that of line level. You will need to make up the gain externally if the mixer will not do it for you.

Microphone level is extremely low in terms of electrical voltages and it is good practise to run your microphone cables over as short a distance as possible. The longer the run, the more noise and interference the cable will pick up which in turn will result in amplified noise on top of your microphone signal.

Microphones are Mono

A mixer designed to accept microphone level will have only one audio input. That will place audio on the left and right channels of the mixer. If you are supplied microphones at line level, you will need to parallel the input between left and right.

Microphone types and quality

Microphones come in two main types of electrical design.

Dynamic microphones have a diaphragm that moves in sympathy with the acoustic pressure waves that hit it. That diaphragm will move a coil through a magnetic field and generate a small electrical voltage, similar to how a dynamo works. This is the mic-level signal that is presented to the mixer.

Capacitor microphones have charged capacitive plates that act as a diaphragm. They move in sympathy with acoustic pressure waves. As one plate moves, the distance between it and the second changes. That in turn changes the capacitive value of the two plates. That value is measured and amplified to produce the mic-level output presented to the mixer. Capacitor microphones require Phantom Power.

Dynamic mics tend to be of a lower audio quality to Capacitor microphones, and for quality studio work Capacitor microphones tend to be the preferred choice.

Phantom Power

Phantom power is a method of providing energy or power to a device along a circuit or set of cables designed principally for something else.

In this instance we are looking at microphones. Dynamic microphones generate their output signal without the need for any external power; however, Capacitor microphones require some power to drive the amplifiers and capacitive fields that generate the output signals. The microphones could have an external power connector or an internal battery; however, by providing power from the mixing desk back along the cable that would normal carry the signal from the microphone to the mixer additional cabling can be avoided.

Phantom power is supplied by applying an identical voltage to both of the two balanced audio connections referenced against the ground connection of the microphone. This is filtered in the microphone to produce power and is also filtered within the mixer to leave just the microphone signal.

Phantom power is normally only available on mixers that have mic-level inputs and is often switch-able between on and off. 48volts is the common provision, although sometimes 15 or 9volts is used. Most mics are designed to work at 48.

The design of phantom power is such that you can plug items that do not require phantom into it, and no harm will come, although there are exceptions and you should be careful. 48 volts is just enough to start causing very small electric shocks, so some care should be taken whilst setting this up.

Microphone Patterns

Microphone patterns describe the functional area around the microphone where it will best pick up sound. For example, some microphones expect you to talk into the front of it. Some will allow a noise in all directions to be picked up. The right pattern is critical to getting the right sound in your studio. Many professional mics have switch-able patterns so that you can choose the type of pattern that best suits your circumstance.

Cardiod:

Allows a specific area in front of the microphone to be picked up, the side area has less pick up and behind the microphone is completely dead. Good for close mic work in noisy environments and will reject some noises coming from other directions. This is a good choice for a stage microphone where rejection of stage noise is important.

Hyper Cardiod:

A very tight cardiod area in front of the microphone will be detected, very good for close mic work and some rejection of noises coming from other directions, also has a small pick up area right behind the microphone. This mic is a good choice if you have several in a studio side by side and need to mic up several people, although be aware of the pick up from the rear.

Omni Directional:

Sound from all directions will be picked up. Used for picking up general atmosphere rather than specific sounds.

Directional/Shotgun:

Used to pick up sounds from some distance away in a specific direction, and will reject all near by sound. Often used to pick up the noise of an event you can not mic up easily, such as a marching band etc. Also used in TV studios on long manually operated boom arms to pick up individual actors or audience members.

Microphone Mounting systems

There are a variety of microphone stands, but the key thing to consider is the actual mounting of the microphone. Most vocal microphones tend to have a rubber band suspension shock mount system. This mechanically isolates the microphone from the rest of the system, so that any vibration on the desk does not get transferred into the microphone as noise.

Handheld microphones can be mounted on a mic-arm with a simple clamp. Handheld microphones tend to be very good at rejecting mechanical noise.

Suggested Microphones for Radio Stations

Studio Microphones for close presenter voice work:

- **Neumann U87:** exceptionally good studio phantom powered vocal microphone, very tight hyper-cardiod pattern, suspension shock mount system, cost £1700
- **AKG 414,** great studio phantom powered vocal microphone sound, very tight cardiod pattern, suspension shock mount system, pattern can be switched, cost £800
- **Audio Technica, AT 4033**, good phantom powered vocal microphone sound, tight pattern, suspension shock mount system, cost £250

News Reporter Microphones for good voice pick up with a little background noise for effect

- **AKG D230,** dynamic, hand-held, low handling noise, nice sound, £110
- **Beyer M58,** dynamic, hand-held, low handling noise, nice sound, £130

Stage Microphone for close voice pick up, rejection of everything else reduces feedback.

- **Shure SM58**, dynamic, hand-held, low handling noise, very tight local pattern, suited to voice, but can be used for most things, exceptionally high rejection of anything not very close to it. Very sturdy and will tolerate artists trying to destroy it. Has been the industry workhorse for 30 years, cost £70

General, all round microphones

- **Beyer 201**, hand-held, dynamic, low handling noise, is suited to all manor of jobs, from drums to voice, cost £150.

There are many other microphones available, but these 7 will suit you through 95% of your broadcasting career.

Digital Microphones

The microphones themselves are still using analogue capacitor charge plates or moving coil diaphragms, however the signals generated are digitised within the body of the microphone, reducing any additional interference or distortion. The microphones will either output AES digital audio directly, or a bespoke standard to a small interface box near by which will convert to AES. The results can be quite stunning.

Remember that microphones are mono devices, and that AES is primarily a stereo standard, the AES feed may need to be mono-ed from left to right, which may or may not be set in the microphone. If the microphone will not do it then it may need to be done at the digital mixer. Check the chosen mixer has this functionality

Microphone Live Lights

It is customary inside and outside of the radio studio to have a warning light, indicating that the microphones are switched on. This is to prevent people from talking or opening the studio door when then they shouldn't and to remind guests and presenters that it's time to perform.

Radio mixing desks should have a "Mic Live" logic output on them, this is normally a contact or switch closure which is automatically operated when a microphone is turned on. It will be wired from the mixer via some interface equipment which in turn will power the "Mic Live" lights inside and outside of the studio.

When choosing the studio mixer check that it has this facility.

CD Players

Sony and Philips introduced the CD standard in the early eighties, and it rapidly took off to become the work horse of music and audio storage. Vinyl is making a slight come back for dance DJ's but in mainstream radio the majority of music played is CD based. CD itself is now rapidly becoming secondary to computer play-out, but it is still a key thing, and is a critical backup to computer based playout.

Professional machines will have most of the following features:

- Balanced analogue outputs
- AES output
- Start/Stop/Cue/Fast Forward and Rewind transport controls
- Remote start/stop/cue controls
- Pitch Shift controls
- Eject lock, to prevent the CD being ejected whilst it is playing
- Single & Continuous Play Mode, allows the machine to be set to play a single track and stop, or play through one track and into the next
- Rack Mounting

All of these are useful, but the one that will make significant difference to operations is the remote start/stop. This is normally an interface connection on the rear of the machine that can be wired to the start/stop buttons on the mixing channel. This will allow the operator to cue the machine up during a quiet moment and then go back to the mixer and use it to fully control the CD player.

Specialist dance mixing CD players will also have additional controls for "scratching" the CD whilst it is playing. This is usually some kind of Jog Wheel that will allow the CD to be speeded up or slowed down or paused. This is an attempt to replicate the world of Vinyl where fingers can be used on the rotating disk to speed up the track or slow it down to match the beats between one track and another. These machines need to be much closer to the DJ or presenter than traditional start/stop CD players.

CD Recorders

CD recorders were designed as production tools or mastering tools rather than studio playback tools. They have a place in the studio but should not be considered for main playback. They are complicated and difficult to use, and is not what most DJs or presenters expect.

Minidisk

When minidisk hit the radio market in the early '90s all of its users fell in love with it, here at last was a digital recording format that was low noise, easy to edit, small and affordable. So small in fact, that many people bought recording minidisk machines for recording interviews out of their own pocket.

The result was an end to cassette tape, and end to portable DAT machines which were unreliable due to the high number of moving parts.

Then in the late '90s something happened, all of the machines that both radio stations and people had bought only a few years earlier all started to fail. They also started to fail at about the same time. The reasons seem to be that any particular minidisk machine has a life of three years max. After which no amount of repair or calibration will make it last much longer. Minidisk as a recording and play-back medium meant that the losses could be quite spectacular. Simply adding an extra few seconds of audio to a disk could result in the complete disk becoming blank.

Very quickly people fell out of love with minidisk, which is a shame, because although purists complain about the digital compression applied to the audio, it is a very neat format. It is physically robust, very small and very light. It failed simply to due unreliability of cheep recording machines.

For the feature lists, it is comparable to CD players, with the added advantage of record and edit.

Professional machines will have most of the following features:

- Balanced analogue input and outputs
- AES input and output
- Start/Stop/Cue/Fast Forward, Rewind and Record transport controls
- Remote start/stop/cue controls
- Pitch Shift controls
- Eject lock, to prevent the Minidisk being ejected whilst it is playing
- Single & Continuous Play Mode, allows the machine to be set to play a single track and stop, or play through one track and into the next
- Rack Mounting
- Track editing and label functions

There are a even few DJ mixing consoles for minidisk that give useful pitch shift and beat mixing functions.

DAT Machine

Digital Audio Tape (DAT) is a good standard with only a few things against it, mainly it is a cassette full of tape moving through a machine. Any machine that has many mechanical moving parts is prone to failure, and DAT is no exception.

It has the same audio quality as CD, full linear digital audio, at a choice of sample rates for record, 32 kHz, 44.1 kHz and 48 kHz. For the high end quality user some versions are available with 96 kHz sampling.

DAT is used very much as a mastering tool for recording, and possibly for playback of long programme items, or even entire programmes. Its response time from pressing play to audio delivery is very poor, maybe a second, so it's not an ideal playback device for regular short items such as music tracks.

As with CD and Minidisk the professional DAT machine will come with a whole host of features

- Balanced analogue input and outputs
- AES input and output
- Start/Stop/Cue/Fast Forward, Rewind and Record transport controls
- Remote start/stop/cue controls
- Pitch Shift controls
- Eject lock, to prevent the DAT being ejected whilst it is playing
- Single & Continuous Play Mode, allows the machine to be set to play a single track and stop, or play through one track and into the next
- Rack Mounting
- Track marking and labelling

DAT is a powerful linear record and playback tool which is industry standard. Some DAT machines have a time code function, where a time stamp is placed on the tape during recording. This is never used in radio, but is used extensively in TV for synchronisation between sound and picture.

Television

Most radio studios have a Television although some don't. The argument against is that the presenters are meant to be concentrating on producing radio programmes and not watching TV. The argument for TV is that the presenter wants to be able to watch 24 hour rolling news to make sure they are in touch with what is happening in the world. TV is an audio visual device and many stations simply have the TV on and turn the volume down. If you are a station that has a heavy news bias, then you may well have broadcast agreements with a number of TV channels. This might allow you to take a press conference live off the TV and put it on the radio.

There are professional televisions available for studios with balanced audio outputs, no speakers and wired remote controls. The use of TV audio in radio is quite rare, so most people use a domestic television taking audio from the SCART socket and converting it to balanced audio before feeding it into the mixer.

There are usually no remote controls for the TV that can be wired out to the mixer, although the infra-red handheld remote will usually suffice.

Outside Source

So far all of the audio sources we have been looking at have been based within the studio. The microphones and CD players are all next to the mixer and the presenter. An outside source (OS) is where the audio is created outside the studio.

If a concert is delivered via a satellite feed, that would be considered to be an outside source (OS), as it is generated and delivered from outside of the studio.

Outside sources often have other associations.

1. They usually have switchers on them
2. They must generate a clean feed.

There may be one OS fader, but there may be several sources generated outside of the studio for example:

- Studio 2
- News booth
- Radio car
- Sat links
- ISDN codecs
- National news provider

These 6 sources would need 6 faders in order to make them all available on the mixer, unless the single OS fader has a switcher or selector to choose one service at a time. It is unlikely that the News Booth and National News Provider will be on air at the same time, so sharing a fader on the mixer seems reasonable.

The OS channel must generate a clean feed.

Clean Feed

When mixing signals to create an output for transmission, it is simply a case of mixing them together and sending that signal to the transmitter. The process is a simple one way communication. There are no other factors to consider.

If you are attempting two-way communications then it becomes a little more complicated. Consider a normal phone call. You need to be able to hear what the other person is saying, and send your own voice back to the person at the other end so they can hear you. If you send the callers voice back to them, they would hear themselves twice, the time they spoke and also the "bounce" back from you to them. This is known as a feedback loop.

A stage show is another good example of feedback loops. The singer on the stage has a microphone to pick up their voice, which is amplified and then fed into a set of speakers to help the audience hear the singer. If the speakers are too loud then the microphone will pick them up and send a copy of the noise from the speakers back to the amplifier, and back to the speakers. This feedback loop can grow and grow until the only thing audible is the hum, whistle or whine of the microphones and speakers in a loop. This is called feedback, and needs to be avoided.

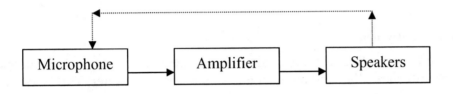

Figure 19 Simple feedback loop

This loop can be seen above and should be avoided as the system will stop responding to the signal you choose to feed it and instead it will start to feed itself and will become out of control.

In broadcasting the principle is the same, if you are trying to connect two studios together so that they can talk to each other, it is important to not send back the signal you have just received otherwise a feedback loop will be created and you will loose control.

It is however necessary for some kind of signal to be sent back to the other studio as they will need to be able to hear or monitor the main studio to interact with it.

To achieve this clean feeds are used where you send back the main programme without (minus) the outside source. This will allow two-way communication, without feedback and you will stay in control.

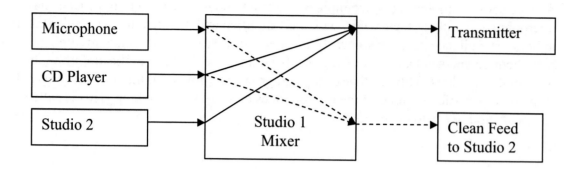

Figure 20 Clean feed generation

In the diagram you can see that all three audio sources, the microphone, CD Player and the feed from Studio 2 are all mixed together and sent to the transmitter. This is known as the programme mix. The Clean Feed to Studio 2 is comprised of a separate mix of only the Microphone and the CD Player. This means that the monitoring feed sent to Studio 2 does not contain the original feed that came from Studio 2 itself.

This will allow Studio 1 to open the Studio 2 fader, take the output from Studio 2 and put it on air, without any risk of creating a feedback loop causing the system to become uncontrollable with the characteristic whistle associated with feedback.

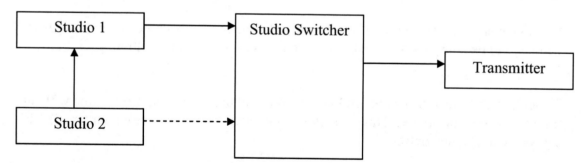

Figure 21 Treating a studio as an outside source

Studio 1 and Studio 2 are able to be switched to the transmitter, but only one can be on air at any one time. Using Outside Sources with Clean Feeds, both studios can be on air and communicate with each other without fear of feedback.

Studio 2 will also have an Outside Source selector, which will create a Clean Feed mix. This is what is sent to Studio 1. In the worst case scenario where both studios fade each other up, there is no chance of feed back.

Telephone Interface

Most stations will want some way for listeners to phone the radio station and interact with the presenters, to take part in competitions or give opinions on daily topics. To achieve this, an interface to the telephone system is required. Because the public telephone system sends and receives its signals on one cable, it is necessary to balance the incoming signal against the outgoing. This is done with an interface known as a Telephone Balance Unit, TBU.

The TBU will convert the single signal into two which can be connected to the mixing desk.

It is still necessary to ensure that the output from your mixer to the TBU is a Clean Feed, a mix that contains the entire studio output apart from the telephone feed itself. This will prevent any kind of echo or feedback being sent from the telephone system back into the studio, and causing strange audio effects.

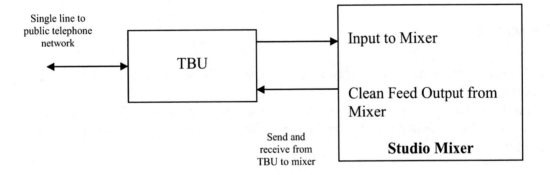

Figure 22 Handling clean feeds for telephone calls

Mixers are designed in different ways, and there are number of methods of dealing with telephone interfaces.

- Use a clean feed
- Use a Telco Channel
- Use an Aux Mix
- Use a mix-minus

The clean feed option we are familiar with, and is the best option where available

The Telco channel is an option that has been added to many mixers which is a channel dedicated to phone interfaces. It generates a clean feed internally on the channel, rather than it being a general mixing desk option, which can be fed to the TBU. The channel also has extra buttons on it for things like, answer, hang-up and talkback to caller.

An Aux Mix is an additional mix bus on the mixer that will allow a second mix in addition to the main programme mix. It tends to be operated from the front user panel of the mixer, and cannot normally be internally set. You can create a "clean feed" using the aux mix, but it does require the mixer operators to understand what they are doing, and will usually cause no end of trouble.

Mix Minus

Mix Minus is very similar to but it is not a clean feed! The logic of its creation is entirely in reverse.

A clean feed is a mix of everything on the mixer, apart from the outside source. It is a separate mix from the main output, created by making two mixes, the main and the clean feed mix.

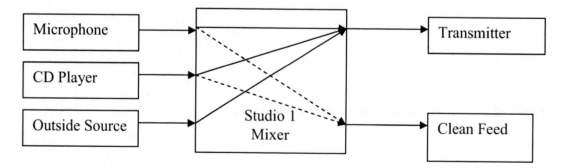

Figure 23 True clean feed generation

A Mix Minus is created by taking the main mix, and electronically subtracting the Outside Source, or TBU, from the main mix, to create an artificial clean feed.

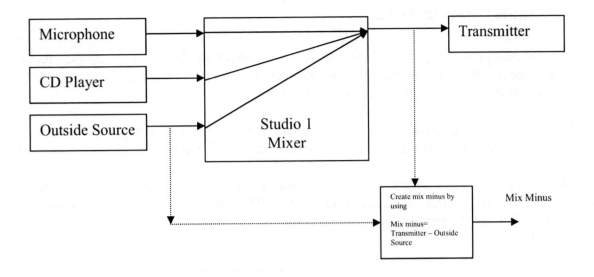

Figure 24 Generating a mix minus

The advantage of a mix minus is that it can be created on demand externally from the mixer therefore you are not constrained by the mixers limitations.

A disadvantage is that mix minus is simply not very good. The external mix minus relies on the level you insert into the mixer remaining unchanged, as it is also the level that will be subtract later. If that level changes the mix minus is not accurate any more and some bleed through the system will occur.

Some mixers provide mix minus internally, that will compensate for any gain changes that occur. These are better, although you will still end up with some bleed-through, due to various harmonics and distortions within the system that cannot be corrected at the final minus stage.

Where you have an option, Clean Feed is always the better technical solution, although mix minus may get you out of a hole if you need it to.

Computer Playout Systems

Studio audio has moved a long way from playing vinyl records and using big old quarter inch tape machines for recording audio and playing it back. Now the world is full of computer based audio storage, editing and playback.

Some radio stations no longer have studios with presenters. The entire radio station output is generated and played out by computer, devoid of human interaction.

We will be looking at the most common usage, which is primary play back within the studio. The studio itself still has CD, DAT and MD for backup should the computer system fail, but the main playback is via computer.

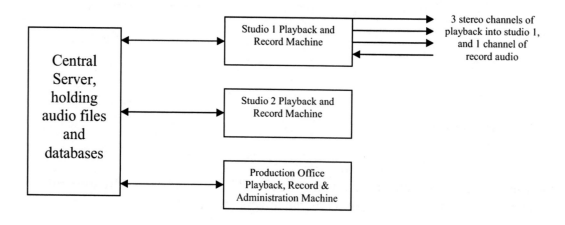

Figure 25 Simple computer play-out system

This configuration has a central file server which holds all of the station audio as computer files. This audio could be music, speech or commercials, and will probably be all three. The file format for the audio can range from linear wave files to MP3, to ADPCM or APTX compressed audio. The preference, given budget and resources, is to keep all audio stored in linear wave files, at a high sample rate, probably that of CD, which is 44.1 kHz, 16 bit. If nothing else linear CD quality sounds better than a compressed file.

The central file server also holds the station databases that are used by the scheduling system to select the music and commercials that are going to be played out on any particular day. Each studio will then have its own playback computer replacing CD or Vinyl playback. So where once you may have had two CD players, you now want two computer playback channels. You will notice we have specified three channels of playback; this is because you will need an additional channel of audio for playback of Jingles.

The studio playout system, should then present you with a running order of material, such as song 1, song 2, jingle 1, song 3 etc… which it will auto-load for you and then just like a CD player, give you the control over when to press start or stop.

In the studio it chronologically may work like this:

Playback channel 1:   Song 1
Playback channel 2:   Song 2
Playback channel 3:   Jingle 1
Playback channel 1:   Song 3
Playback channel 3:   Jingle 2
Playback channel 2:   Song 4
Etc.

Playback channels one and two act as the old traditional CD players, swapping from one to the other and then any jingles play from the third channel.

The studio computer system can also record audio from the mixer, in the on-air studio. This will only be occasional, as the main job of the studio is playback and not record. The second off-air studio or a production machine may be used for recording more often. This is how new songs, voice links or commercials will be loaded onto the system.

It is also possible for audio, in an existing computer format, to be auto-loaded into the system. This allows remote agencies that produce news, or programmes, to load audio directly into the computer ready for play out, without needing to take up valuable studio time.

Computer systems can work in either the analogue or digital audio domain, and choice of soundcard will dictate how easy or not that is.

There are many play-out systems on the market, and you will need to find one that suits your needs, questions to ask when looking include:

- Will it work with linear audio
- How much audio can it store
- How much will it cost
- How many studios will it work with
- How is it licensed, per system or per studio, or by audience size
- How does it load audio
- Can it rip CDs directly
- Will it import audio files
- Will it normalise audio levels
- Will it work in a Live Assist mode, to help the presenter
- Will it work in an automation mode to replace the presenter
- Will it allow voice tracking to record the presenter for playback later
- How does it import traffic/commercial logs for play-out
- How does it import music logs for play-out
- How often does it need to be shut down for maintenance
- How do you back up all of the data
- What happens when the main server fails
- How many audio files will it play at the same time in a studio
- How do you start and stop the audio, can the mixer start stop buttons be used
- How do you pre-fade a track before play-out
- Can you edit segues in the studio before they are played out
- Does it offer log files to find out what went wrong
- Does it offer commercial reconciliation files
- How long does it take between loading audio and being able to play it out
- Does it have a built in audio editor for production
- Can it be remotely controlled from home to resolve problems at weekends
- Is it simple enough to be operated by presenters

By asking these questions and getting answers that seem reasonable you will be well on the way to getting a computer play out system that will suit your needs

Monitoring

A broadcast mixer will give you several monitoring options:

- Studio 1 output
- Studio 1 Pre-Fade
- Studio 2 output
- Transmission output
- Off Air receive

It would be reasonable in Studio 1 to want to listen to the main programme output that you are generating, i.e. Studio 1 output.

It would be useful before putting a CD on air to be able to listen to it first, just in case there is a problem with the content, or the audio level. This mixer function is called Pre-Fade, and it allows you to listen before the faders opened and the item is committed to air. The pre-fade function is found on each channel so that you can select what you want to listen to The mixer itself will have an overall listen to pre-fade option, so you decide if it comes through the headphones, or through the main speakers.

Listening to other studios is also useful, this will allow you to monitor what is happening elsewhere in the radio station.

Listening to transmission is very useful. It lets you check what you are sending to the transmitter. If you have a fault somewhere that has taken you off-air, you should check that you're actually sending audio to the transmitter. This will help locate any faults to the studio, the station or the transmitter.

Off-air monitoring is the ultimate proof that you are on-air. If you are generating a radio programme in a studio, sending it to the transmitter and you are listening off-air, using a genuine radio to pick up your broadcasts, you can be sure you are on air.

New digital links and audio processing technologies are starting to add time delays between the studio and the listener. In some cases it is not possible for the presenter in the studio to listen to what is being broadcast because of these delays. An effect is created that is like talking to an echo.

The option to listen off-air is still important, as the presenters should check every now and again to ensure that they are actually on-air.

Studio Layout

The studio layout is a key part of the station design, this is where the presenters and producers are going to work for the years to come and to change the basic layout once it has been installed is costly and time consuming. Understanding the role of the studio will go a long way to creating good design.

Most studios are designed for a single operator to use, with provision made for occasional guests. Some studios may need to be designed with multiple presenters in mind to suit programmes such as the breakfast show "zoo" format, with many presenters and guests all needing to be on-air with access to microphones at the same time.

Consideration also needs to be made for other regular contributors such as news readers. Will news be read live in the studio, or from the newsroom or a news booth? If it's from the studio there needs to be a good position with easy access to microphone and headphones, as well as space for paper scripts or access to a computer terminal from which to read the news scripts. Audio clips may also be required in the news and provision must be made for either the presenter or news-reader to do this.

Once the studio operation is understood the layout can be designed. Forward planning also needs to be considered, will the design cater for potential changes over the coming years? Is it likely extra producers, presenters or operators will be employed. Will the single presenter be replaced by a more complex "zoo" format? With these issues in mind the design can be made to cater for all needs in the future. Unexpected changes will be expensive to deal with later.

The location of the studios within the building will also affect the design

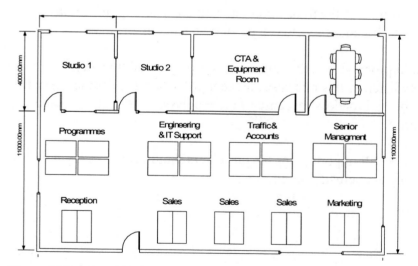

Figure 26 Layout of PopFM

We know that Studio 1 has three windows, one to the outside world, one into the office and one looking into Studio 2. There is also a door. That leaves one wall free.

Should the design allow for a presenter in Studio 1 to look at the presenter in Studio 2? If one studio is regularly going to hand over to the other for a programme change then maybe they should be able to see each other. If the presenter is going to face the main outside window will there be an issue of sunlight blinding anyone in the room, or if the presenter has his back to the window will sunlight cause glare on the computer monitors making them unreadable. Or is the simple solution to fit a blind?

This studio layout is suggested for a single presenter self operated studio that will never be able to accommodate any guests

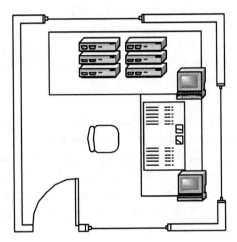

Figure 27 Simple self-op studio layout

The presenter can see through all three windows without having to move away from the mixer position. This enables a line of sight conversation with a guest or presenter in another studio. The presenter can also see through the window into the office, and has daylight presented to him side on, which in most cases should not cause too much glare onto the computer screens.

The computer screens are laid out one left and one right of the main work position and all of the extra equipment such as CD players and DAT machines are rack mounted away to the left hand side and will be operated with remote controls on the mixer.

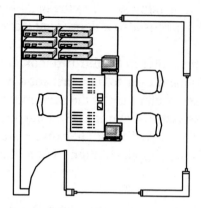

Figure 28 Self-op studio layout with guest positions

This studio layout uses the same physical space, but everything is much more compact to squeeze more functionality into the space. It also offers the advantage of fitting two guest positions into the studio, to allow interviews, or news reading to take place. It will even allow for a double headed (two presenter) breakfast show or for a programme to have a technical operator driving the studio whilst the main presenter sits on the guest side of the desk.

This is a much more traditional design style, although it could still present problems. If the studio is mainly one single operator/presenter then the foot traffic route into the studio is simple, open the door and walk directly to the chair with no obstructions. If there are going to be many guests then the route is more difficult, they must open the door to enter the studio then allow it to close before they can more round to the guest side of the desk. If this is to be a rare occurrence then this won't be a problem. If it's a regular occurrence then it is.

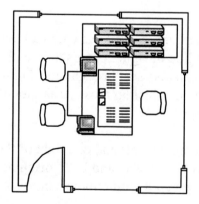

Figure 29 Reversed self-op studio layout with guest positions

This layout favors the guests, making their access to the studio space easier, although the main presenter or operator will have more difficult access.

The problem of the door blocking access to a large part of the room may simply be resolved by hanging the door to swing open to the left rather than the right.

If it is important that the studio be attractive either for the guests or staff using it, or for media coverage over the coming years, then aesthetic design may also be a consideration.

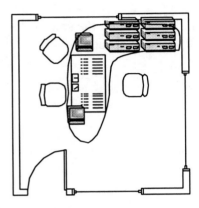

Figure 30 Studio layout with curved design

Studios can be built using curves rather than straight edges, the key things to remember being they cost more, and will take longer to make.

"Attractive" studio shapes may also offer problems with the technical equipment, for cable management or space to mount equipment in. Again this can be resolved with careful planning, and if necessary building a mock up from disposable materials first.

With radical designs mock ups may be crucial in showing all kind of problems, structural, physical and technical. Time spent checking out studio layout and mocking up designs will result in fewer long term problems and lower costs.

Where to put the speakers

The floor to ceiling height of the studio is the leading factor to consider when placing the studio monitor speakers.

Ideally they will be placed left and right of the mixer/operator position and probably hung from the ceiling. This will allow them to be out of the way, and high up enough for no one to hit their head on them, and for them to be angled down toward the user.

If the studio ceiling height is low it may be necessary to place them on the studio woodwork, although the disadvantage here is they take up a lot of space.

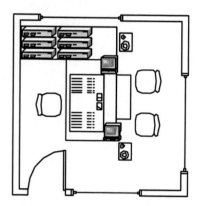

Figure 31 Studio monitor locations

Monitors close to the operator need to be "Near Field Monitors" which are designed for the optimum listening position to be near the speakers.

Far field monitors are designed for the best listening position to be some distance away, if for example the monitors had to be placed on a far wall.

When designing the studio fabric it is important to consider what might be hung from the ceiling. This will ensure that the structure can support the weight of anything that may need to be installed. Anything hanging over a position where someone may be stood, or sat below, it also needs to be protected by an additional safety chain or restraint. This will ensure that should the main support fail, the overhead item does not come crashing down on top of someone.

If the station regularly has famous people visiting, imagine the law suit for damages or the adverse image the company should suffer if a speaker fell out of the ceiling and injured them.

Should presenters sit or stand?

The debate over presenters sitting or standing is still on going and will be for many years. There is little doubt that a station that makes its presenters stand, benefits from a more authorative presentation style.

Sitting causes the diaphragm within the body to be compressed and not perform to its best. Standing, however allows the diaphragm to fully perform and the voice is given authority and power. Few people doubt that standing gives a better "sound", however, many people can't imagine forcing their presenters to stand for 3 or 4 hours per day.

The reality is that the type of programme or style the station is trying to deliver will have a lot to do with the decision making. If the station is looking for a fast or authorative programme then maybe standing will work, if however you are looking for a relaxed warm sound then sitting will suit just fine.

The technicalities are simple, other than to say, once the height is chosen, it is very difficult to change that later. Without significant cost implications to completely rewire the studio the height is fixed from day one for the life of the studio.

The two standard heights are 72-78cm for sit down, and 100cm for standing, measured from the ground to the table top.

As with any station, this can change by a few centimetres in any direction.

Patress Panel

Within the line of sight of the main presenter or studio operator should be some method of passing critical information from the studio systems to the operator. A patress panel is good way of doing this.

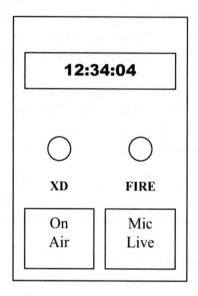

Figure 32 Studio patress panel

This panel will normally be placed on the wall opposite the presenter and contain useful information such as

On Air:	Indicator that the studio you are in is on air, or not
Mic Live:	Indicator that the microphone in the studio is live
XD:	The studio X-Directory phone line is flashing
FIRE:	A silent visual indicator that the fire alarm has been activated
Clock:	Time

It is important when building the studio to work out the best place for this, and also to cable in advance so that the panel can easily be installed. It will require; mains power, signal cable, and also sufficient depth in the wall to take the panel of about 100mm.

Power Distribution

Each physical device in the studio will need to have a power supply. A quick count suggests the following devices will need power.

1. CD 1
2. CD 2
3. MD 1
4. DAT 1
5. TV
6. Telephone
7. Computer Play out Screen
8. The mixer
9. Amplifier for monitor speakers
10. Spare

Rather than having ten power outlets round the studio, there should be a single technical power feed brought in from the CTA and presented to a mains distribution unit (MDU) within the studio. This is a box designed to distribute power to many devices and has neat cable management and safely fused outputs available. If running a dual mains system then use a dual input MDU which will switch between primary and back up power.

Using an MDU will help isolate the studio in the event of any problem. One switch that controls everything is much safer than tangled cables everywhere.

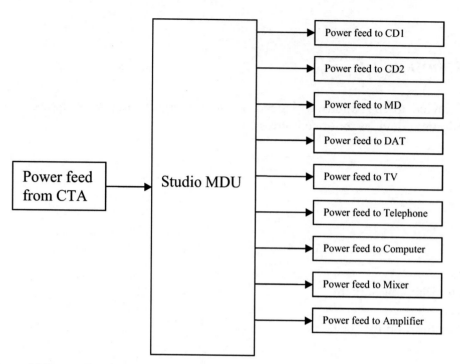

Figure 33 Power distribution within studio

With a neatly installed MDU and power cabling that is labelled, a single item of kit can be identified, turned off and worked upon. There are also options for dual fed MDUs to provide better power availability and also intelligent MDUs which are able to connect to computer networks and inform you of any problems they experience or be remotely started or shut down.

Once you have calculated the studio power requirements, you may wish to install additional infrastructure for any future developments. Putting in additional power handling at this stage means that any additional equipment installed can be done neatly without major disruption and cost.

Flexible Installation Techniques

Studios can be built and designed to be flexible for the users, but no matter how much careful planning goes into the design stage, something will crop up on a weekly basis that was never considered, or finances would not allow for, in the planning stage. There are several ways to deal with this; ignore it, rebuild the studio each time a request arrives, or use a flexible patching and installation system that allows services to be connected, disconnected and changed in seconds.

There are two main systems used in both analogue and digital audio installations that add flexibility and allow regular and irregular changes to be accommodated. They are frequently used in station installations. "Jack Fields" allow frequent short term audio patching for both the advanced studio user and also the engineer. It is flexible and allows short term audio patches and changes to normal service.

"Krone Fields" allow flexible installation of equipment, which can be changed for long term reasons with some ease. This is designed for use by engineer rather than user and changes made are either long term or permanent.

There are also computer services within a studio to consider, the audio play out machines are often kept outside of the studio location to cut down in-studio noise. Keyboards and screens will need extending from the CTA to the studio location. This can be done using dedicated long cables, although these tend to suffer from noise and degradation over the distances involved and are not very flexible. CAT5 computer patching allows for standard extenders to be patched from area to area and is very flexible.

It is critical with all of these systems that documentation is made at the design stage, and with Krone in particular, whenever a change may take place. Whilst each individual item is simple, the overall volume of these patches means that without a clear documented map of the system that it will immediately become outdated and unwieldy.

Jack Field

Jack Field is a flexible patching solution that will make short term modifications easy to manage.

The idea is that each item of equipment is connected to a patch panel rather than directly to the mixer. The patch panel in its normal state of standard design automatically connects each item of kit to its designated part of the mixer. When the requirements change, cables are inserted into the Jack Field patch panel to re-configure the mixer and equipment operations.

An example of how this works would be a small radio or HiFi. In its normal operation the music comes out of the HiFi speakers, however, by inserting a set of headphones into the headphone jack socket the sound stops coming out of the speakers and starts coming out of the headphones. You have "jacked-in" the alternate headphone audio destination to the system.

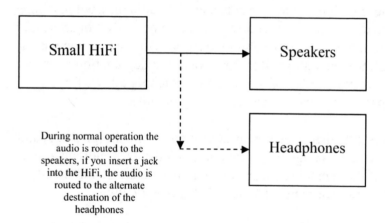

Figure 34 Example of alternate audio routing

The same is true in radio station Jack Field, the normal operation for a CD player is to be connected to the CD fader on the mixer, however by over-patching you can make the CD appear on a different fader or you can insert new audio on the CD fader.

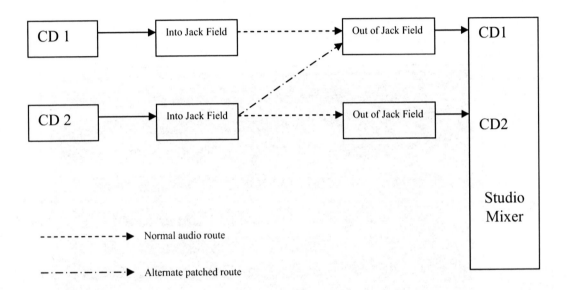

Figure 35 Basic Jack Field example

In this configuration CD1 is wired via the Jack Field into the normal CD1 input of the mixer. CD2 is wired similarly into the CD2 input of the Jack Field. Using the patch field we can take the output of CD2 and over patch it into the input for CD1. This means that when CD2 is playing the audio will appear on the CD1 fader.

Jack Field is flexible and depending on its configuration can be even more versatile. For example the audio from CD 2 could now appear on the CD1 and CD2 faders, or the audio on the CD1 fader could be a passive mix of both CD1 and CD2. Alternately a third entirely new item of kit could be plugged into the Jack Field, such as a TV or record player. This level of flexibility allows you to deal with unpredictable situations without having to completely dismantle and rewire your radio studio.

Jack Field will often have all the main studio audio sources and destinations made available to it. Lay them out in a simple and straight forward manor for easy understanding and use. Start with sources and their expected destinations, for the input side to the mixer. Then work with the mixer outputs, which become audio sources to their destinations.

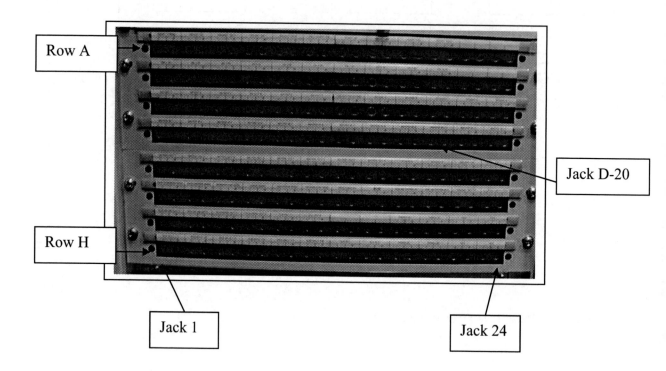

Figure 36 Navigating a Jack Field

		1	2	3	4	5	6	7	8	9	10	11	12	13	14	15	16	17	18	19	20	21	22	23	24
Listen	A	Mic 1L	Mic 1R	Mic 2L	Mic 2R	CD1 L	CD1 R	CD2 L	CD2 R	MD L	MD R	DAT L	DAT R	TV L	TV R	OS L	OS R	TEL L	TEL R	PC 1L	PC 1R	PC 2L	PC 2R	PC 3L	PC 3R
Break	B	In 1L	In 1R	In 2L	In 2R	In 3L	in 3R	in 4L	in 4R	In 5L	In5 R	In 6L	In6 R	In 7L	In 7R	In 8L	In 8R	In 9L	In 9R	in 10L	in 10R	In 11L	In 11R	In 12L	in 12R
Listen	C	Mix Out L	Mix Out R	LS L	LS R	Cans L	Cans R	Out 1L	Out 1R	Out 2L	Out 2R	Out 3L	Out 3R	Tel out L	Tel out R	OS CFL	OS CFR	Aux Out L	Aux Out R	spare	spare	CTA tie1	CTA tie2	CTA tie3	CTA tie4
Break	D	TX ST1 L	TX ST1 R	Amp L	Amp R	Mon 1L	Mon 1R	MD inL	MD inR	DAT inL	DAT inR	PC Rec L	PC Rec R	Tel in L	Tel in R	OS CFL	OS CFR	spare	spare	spare	spare	STD 2 tie1	STD 2 tie2	STD 2 tie3	STD 2 tie4

Figure 37 Jack Field layout for PopFM studio 1

The top row (A) shows all the audio sources, items that generate an audio signal. For example A11 and A12 are the audio outputs from the DAT machine. The second row shows all the mixer inputs. B11&B12 are the mixer inputs for channel 6, left and right.

This Jack Field is configured as "listen/break". Row A being listen to sources, and in normal usage with no jacks inserted is internally wired so that audio is directly connected to the jacks on Row B. So the DAT outputs on A11&A12 are connected directly to the mixer inputs 6 L&R.

This allows you to insert jacks into the "break" rows with an alternate audio source. For example if you inserted a pair of jack leads going from the minidisk outputs (A9&A10) to the mixer inputs 6 (B11&B12), then the audio from the minidisk would appear on both the minidisk fader and also on the DAT fader. The DAT machine output would no longer be connected to the mixer and would not be available. This would require two "patches" a jack from A9 to B11, and another from A10 to B12.

Rows C&D are configured similarly in the listen/break format. Row C contains the audio outputs of the mixer, main output for the studio C1&C2 to be sent to the CTA on D1&D2, the local loudspeaker and headphone outputs for the studio, and then a series of additional outputs 1, 2 and 3. These are additional record outputs separately generated by the mixer for feeding to DAT and Minidisk type devices.

The principle still stands that if, for example, you wished to take a CD1 output and record it directly to Minidisk, you could patch from A5 to D7 (left) and A6 to D8 (right). Because the A row is listen, you can parallel the output with a jack lead and the break the normal Minidisk input away from Out1 and insert the jacks into D7&D8 to replace the normal audio source with the CD output.

This could be useful if you needed to copy a CD to Minidisk whilst using the studio for something else, such a production session, or simply being on air.

Rows C17 to 24 and D17 to 24 are configured slightly differently. They have been wired as listen only. C17 & C18 are Aux outputs from the mixer that have no designated destination. They do not need to go to an output and so have no need to be listen/break, only listen.

C21 to 24 are tie lines from the studio to CTA, again not needing a listen/break configuration. They can be used if you have additional equipment in the CTA that needs to be wired into the studio without redesigning the entire station. Equally they could be used the other way round; you may for example wish to patch CD1 from studio 1 to the CTA. Ties have also been provided from Studio 1 to Studio 2, again there may be a source of audio in either of these studios that you need to provide to the other.

Jack Field is all about flexibility and making it easy to move audio from one place to another, the more you have in all locations the less major reconfiguration work you will need to do in the future.

Rear
Wiring of
Jack A-24

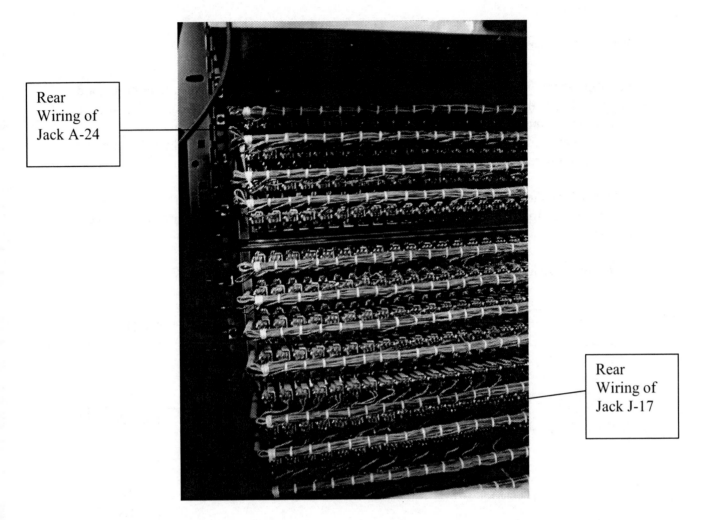

Rear
Wiring of
Jack J-17

Figure 38 Rear construction of Jack Field

The construction of Jack Field is simple but considerable. It is literally a metal plate with many hundreds of jack sockets mounted on it. In turn each of those sockets needs to be wired to the audio source or destination. In the case of the listen/break configuration such as the A1/B1 jacks, they also need to be wired together directly on the back of the panel. This in itself is a major task and not one you will want to change very often. If for example you decide not to use Minidisk anymore and wish to replace it with a new device you do not really want to re-wire the complex Jack Field. Instead there is another more flexible solution that can sit between the audio device and the Jack Field, a Krone Frame.

Krone Frame

Flexibility is the key to all technical design, including radio stations. No matter how good the planning or the brief, a time will come where re-configuration will take place. New technologies emerge that are quickly adopted, new working practices change the requirements and, in some cases, poor initial choices simply need fixing. In order to make the change process as painless as possible, good flexible design is needed.

Whilst a Jack Field allows a quick change to the immediate working environment, re-routing audio from one place to another, the key to a more permanent change is a Krone Frame.

So far we have looked at how the Jack Field sits between two items of kit.

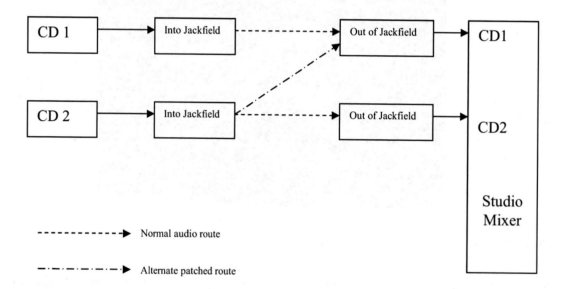

Figure 39 Simple Jack Field audio route

Each item of kit, be it CD player or mixing desk is presented to the Jack Field to allow for maximum operational flexibility. We now need to look at technical flexibility. Each jack and item of kit must be wired directly to the Jack Field. If an item of kit moves becomes obsolete or a reconfiguration is required, then the cabling to the Jack Field has to be replaced.

The cabling on the back of a Jack Field does not lend itself to easy or regular working. If you need to work on the station Jack Field, whilst programmes are being generated and passing through it, then it moves from complicated to risky.

A mid section of connectivity is required, a Krone Frame. All items in a technical area, (studio or CTA) are wired first to the Krone Frame and then to the Jack Field. This means that delicate items such as the Jack Field can remain undisturbed whilst the long term configuration of the station or studio can be done on the Krone Frame, which is better suited to being worked upon.

Figure 40 A single krone block

Similar to the Jack Field theory, all items of kit are wired to the Krone Frame and then patched together to produce a working audio system.

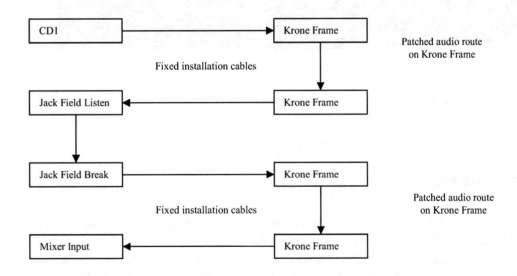

Figure 41 Example of an audio path via a krone block

Figure 42 A large Krone Frame in use

Use of a krone block

Each block consists of 20 pairs of connections, 10 input pairs which are presented to the top of the block, and 10 output pairs connected at the bottom of the block. In normal operation the inputs are connected directly to the outputs. Similarly to the Jack Field, these direct connections are via sprung contacts that are normally closed and can be broken by inserting a break tool or connector. The break options allow circuits to be disconnected or paralleled for measurement whilst a circuit is in use, or alternate circuits to be inserted.

The Krone block

In the diagram below you can see each pair is presented numbered 1 through 0. Installation cables are presented to the top of the block, and operational patching (or "kroned" connections) is via the bottom of the block.

Whilst most circuits should be presented via a Jack Field, the following example will ignore it for clarity.

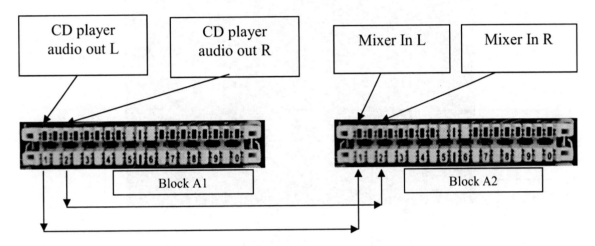

Figure 43 Patching an audio route between krone blocks

In this example we have used the top side of Block A1 for the permanent installation of a CD player, and the top side of Block A2 for the permanent installation of a studio mixer. It is most likely in years to come that you will still have CD players and mixers. The way they are used might change. In this example we have patched the bottom side of Block A1 pair 1, to the bottom side of Block A2 pair 1 and the bottom side of Block A1 pair 2 to the bottom side of Block A2 pair 2. This provides an installed audio path of CD player to studio mixer.

In krone jargon this patch may well be described as "jumpering" A1-1 to A2-1, and A1-2 to A2-2.

This patching is easy to do, and only takes a few seconds. The installation of new direct cabling from the mixer to CD player, should a change be made, could be quite a time consuming job.

Navigating a Krone Frame

A large krone frame may contain thousands of blocks, each containing up to 20 pairs of audio cable connections. A way of navigating around the frame is required.

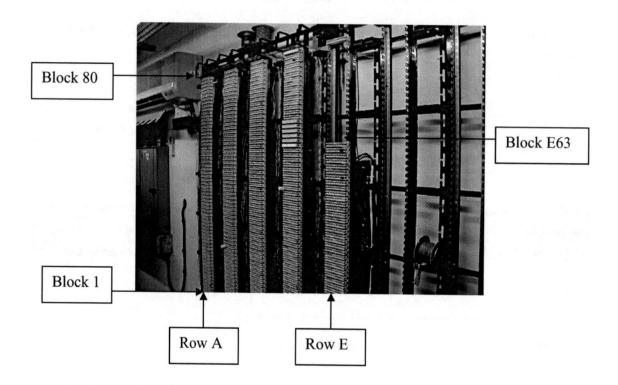

Figure 44 Navigating a Krone Frame

The bottom block nearest the floor is known as Block 1, and the second Block 2.

The block at the very top of the left hand side will be known as Block A80. Block E63 demonstrates 5 rows along from the left hand side, and 63 blocks up from the bottom.

The very bottom block on each row A1,B1,C1,D1 etc is often an earth block, which allows for earth termination of any DC signalling or cable screens.

Krone blocks come in a number of styles, the standard Krone 237A which is the normal disconnect block for audio jumpering (patching) is yellow. They are also available in other colours such as blue and grey that can be used to signify DC paths. Red blocks are also available to signify earth bars which are all fully paralleled. This allows the block to be wired to the star-point earth and then used as a local earth reference.

The Krone Frame is designed to allow installation cables to be brought in from above or below the frame, behind the main mass of blocks. Individual cable pairs are brought in from behind onto the top side of the blocks. Jumpering is then performed on the bottom side of the block on the front of the frame and loomed neatly through the side cable guides into the vertical gaps between the blocks. These vertical gaps will then act as guides for the patch, or jumpering, cable as it navigates its way round the frame. Often going up to the top of the frame, across the top guides and then back down another vertical guide to the destination block.

The bottom patch side of each block is designed to take one pair of patch cables for point to point connection. It is common practice for a circuit that needs to be paralleled or split for a second set of jumpers to be applied directly on top.

It is also common practice to use different coloured jumper cable, for example, you could use red and white for analogue audio, red and blue for AES digital audio and single blue for DC patching.

Krone Map

The Krone system is powerful and can be complex. Detailed documentation of all work should be made, including any changes that are made over the life of the system. A summery map should be made with detailed records of each block.

Block	Row A	Row B	Row C
15			ST1-St2 ties 11-20
14			St1-St2 ties 1-10
13		JF D21-24	CTA ties 21-30
12		JF D11-20	CTA ties 11-20
11		JF D1-10	CTA ties 1-10
10		JF C21-24	
9		JF C11-20	
8		JF C1-10	
7	Computer	JF B21-24	Mixer DC
6	TV OS TEL	JF B11-20	Mixer Out 2
5	DAT	JF B1-10	Mixer Out 1
4	MD	JF A 21-24	Mixer in 11-12
3	CD 1 & 2	JF A11-20	Mixer in 6-10
2	Mic 1,2	JF A1-10	Mixer in 1-5
1	Earth	Earth	Earth

Figure 45 Studio 1 PopFM Krone Map

The map shows the basic layout of the Krone Field within the PopFM studio 1. From this we can quickly see all the key components of the studio. Row A on the left hand side with all of the outboard equipment such as CD players. Row B shows the entire Jack Field, and Row C shows the connections into the mixer and also the ties from the studio to the CTA and also to Studio 2.

Krone Records

Once the master map has been generated a detailed description of each block is also required. This will form the master jumper record where you can trace an audio route from a CD player via the Jack Field into the mixer and then ultimately to the CTA for transmission.

A single krone record sheet may look as follows

A3	Studio 1 CD1 & CD2 Audio Outputs L&R			
Pair	**Description**	**Jumper 1**	**Jumper 2**	**Notes**
1	CD1 L	C2-5		Direct to Mixer in 3 L
2	CD1 R	C2-6		Direct to Mixer in 3 R
3	CD1 AES			
4	CD1 start/stop			
5	CD2 L	C2-7		Direct to Mixer in 4 L
6	CD2 R	C2-8		Direct to Mixer in 4 R
7	CD2 AES			
8	CD2 start/stop			
9				
0				

Figure 46 Example of a single krone record sheet

By looking at the main krone map for Studio 1 the CD audio outputs can be found on Block A3. The record sheet for Block A3 shows the CD1 audio outputs left and right can be found on pairs 1 and 2. In this simplified case they have been patched directly as the record shows to the inputs of the mixer on block C2 pairs 5 and 6.

Looking at Block C2 the mixer inputs are shown. Pairs 5&6 are the Left and Right direct inputs to mixer channel 3 and it has been nominated as the CD1 input. Just as Block A3 1&2 claims to be patched to C2 5&6, it can be seen on Block C2 that it claims C2 5&6 are patched to A3 1&2.

C2	Studio 1 Mixer Inputs 1-5			
Pair	Description	Jumper 1	Jumper 2	Notes
1	Mixer in L1	A2-1	C2-2	From Mic1 Amp
2	Mixer in R1	C2-1		Parallel Mic 1 Amp
3	Mixer in L2	A2-2	C2-4	From Mic2 Amp
4	Mixer in R2	C2-3		Parallel Mic 2 Amp
5	Mixer in L3	A3-1		Direct from CD1 out L
6	Mixer in R3	A3-2		Direct from CD1 out R
7	Mixer in L4	A3-5		Direct from CD2 out L
8	Mixer in R4	A3-6		Direct from CD2 out R
9	Mixer in L5	A4-1		Direct from MD1 out L
0	Mixer in R5	A4-2		Direct from MD1 out R

Figure 47 A second krone record sheet

Block C2 also shows the other inputs for the mixer on channels 1 to 5, this tallies with the krone map shown. This is the method used for documenting the krone layouts within a studio build.

The jack field interface has been ignored for this demonstration; however the patch route for CD1 to the mixer through the jack field may read,

Left channel CD audio to JF to Mixer A3-1 to B2-5, B5-5 to C2-5

Right channel CD audio to JF to Mixer A3-2 to B2-6, B5-6 to C2-6

The connection from B2-5 to B5-5 being made within the internal jack field normaling for the left hand channel and B2-6 to B5-6 is the same.

Working with Krone Blocks

Building a krone frame and preparing the cables is the key to the making an installation flexible.

Firstly you need to prepare the cable to apply to the krone block. This is the fixed side of the installation, that maybe a tie between two areas or the direct cabling to a CD player or mixing desk. There are many cable standards that work with krone systems, here we will be concentrating on PSN cable, which is primarily used for inter area ties, or cabling to large connectors with more than just one pair of audio presented. The principles still stand for smaller installations.

Peel back 20 to 30cm of insulation from the PSN cable to reveal the inner cable pairs.

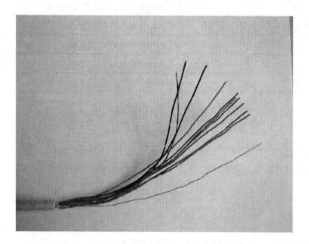

Figure 48 PSN multi-core cable with outer insulation removed

This cable contains 5 pairs of audio cables and a single screen wire, sometimes known as the drain.

The cables are colour coded

Blue/Blue-White is the 1st pair
Orange/Orange-White is the 2nd pair
Green/Green-White is the 3rd pair
Brown/Brown-White is the 4th pair
Slate/Slate-White is the 5th pair.

The Blue on the first pair is a solid colour, and the Blue-White is a solid blue colour with White bands or stripes painted onto it. The Blue is known as the Phase or Hot cable and the Blue-White is known as the Anti-Phase or Cold cable, this reflects the balanced audio standards.

There are PSN cables that can contain as many as 50 pairs of audio cable, and they use this same basic colour code, but start to change the white stripe for other colours to identify which pair in the cable they are.

Next prepare the end of the cable before it is applied to the krone block.

Figure 49 PSN prepared ready for application to a krone block

Twist each colour pair starting with Blue/Blue-White, together to create 5 pairs of cable. Then cut the solid screen, or drain wire, and solder a green (earth) wire on it, which will act as the link down to the earth block at the bottom of the Krone Frame. This will provide the earth reference for the cables screen. To avoid earth loops remember only to connect the earth at one end.

Next add a sleeve or heat shrink to the area where the insulation ends and the main cable begins, this will act as a strain relief to the cables and prevent them from snapping under stress or catching on rough edges. It will also cover and insulate the screen you have just soldered on. Present the cable through the rear access of the krone block from below.

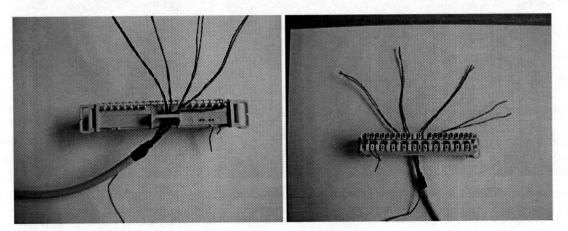

Figure 50 A prepared PSN cable being applied to a krone block

The sleeved cable enters the krones cable management from below and behind. Then peel the first blue/blue-white pair off into the first position for termination. The right hand image shows them from the front.

Using a krone insertion tool, punch the blue/blue-white cable into the upper positions above position 1 on the block. Note that the solid blue or the hot, goes to the left of the numbered position, and the blue-white goes to the right of the position. Continue to populate the positions along the block according the cable colour number sequence.

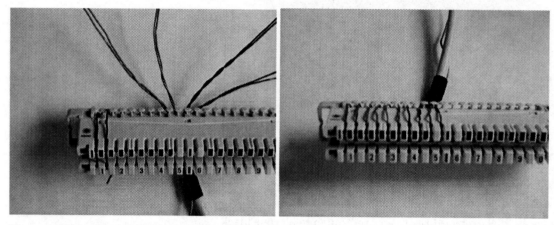

Figure 51 Terminating PSN onto a krone block

The block now has the full 5 pairs of the audio cable terminated on the top side of the block. The cable should then be cable-tied around the rear access point to prevent any movement and to offer further strain relief.

The blocks at this stage will be mounted up on the krone frame, and patching from one block to another will then take place. The frame itself contains cable guides for the patching cables.

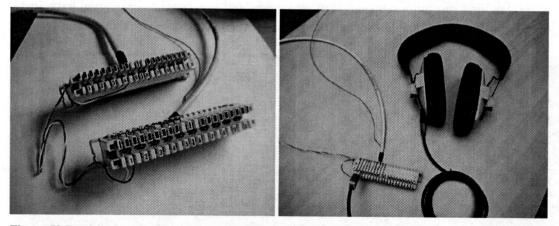

Figure 52 Patching a signal between krone blocks, and then monitoring it

The picture on the left shows the upper block being patched from pair 4 to the lower blocks pair 1. Note how the jumper cable goes via the side cable management point.

Good practice suggests that any jumpering on pairs 1 through 5 go via the left hand cable management and pairs 6 through 0 via the right hand cable management.

One of the advantages of analogue audio in krone systems is that you can monitor a circuit by simply plugging in a pair of headphones without significant adverse effects on the signal path.

Krone Conventions

There are many cabling conventions and krone systems are no different. Conventions also change, depending on where the system is and who installed it. Once a convention is set it is often best to follow the local one rather than try and retro fix it, unless of course it is very wrong.

Jumpering within the krone system is one such area, traditionally jumpers on pairs 1 through 5 pass through the left hand cable management system and jumpering on pairs 6 through 0 got the right hand side. More recently installation practise has started to use only the left hand side, making all pairs of jumpering 1 through 0 pass to the left. This does make for a clean and tidy looking install, but can make the cable mass passing through the management hole difficult.

There is also a tendency when jumpering between rows to pass all of the jumper cable through the bottom access points on the frame itself, this is usually because its easier to reach. It can lead to the bottom cable management becoming very congested and so prudent early patching through the top side of the frame may help.

When jumpering there is a delicate balance to be made between the cables being tight enough between blocks to stay rigid so as not flop about and also being slack enough for them to be traced. Tracing an undocumented cable can be easier if it can be pulled, watch the frame to see if the other end moves, if it doesn't it will be a hard job to find it, especially if the frame is densely populated.

There are tracing tools available, where you insert a krone plug, similar to the one shown with the headphones attached, and inject a radio frequency carrier. Then using a detector the krone frame can be scanned to search for the path the cable system takes. This can be invaluable if working on an old undocumented frame. The radio frequency does not interfere with analogue audio and so the system, with caution, can be used on live transmission circuits.

When doing any work on a live transmission system it is prudent to listen off-air to a radio, so should you inadvertently jumper incorrectly or remove an incorrect cable you can hear it on the radio and put it back very quickly.

CAT 5 Computer Patching

CAT 5 is a well documented standard within the IT world, and other publications will be better placed to describe it. This section will examine the best way to exploit it for broadcast purposes. In much the same way that audio installations can be made flexible, so can computer installations.

It is advisable to install studio computers outside of the studio for the simple reason they make a lot of noise, mostly through their cooling fans. This means that the keyboards, mice, screens, serial and sound will need extending into the studio where the machines operation is required.

The common solution is to use a CAT5 Keyboard Video Mouse (KVM) extender. This is a neat box that connects to the KVM ports on the back of a PC and then presents the user with a CAT5 socket. Using a CAT5 patch lead to a matching KVM receiver, a new set of Keyboard, Video and Mouse ports are presented, in this case in the studio.

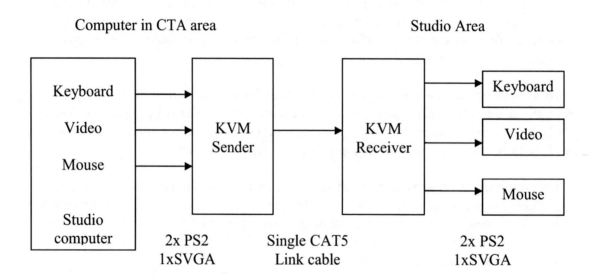

Figure 53 Keyboard Video and Mouse extended from a computer to a studio

The PS2 & SVGA cables are designed to be connected locally to the machine they are controlling and work well for maybe 5 metres before noise and signal degradation effects take place. The KVM extender systems allow the keyboards video and mouse to be many hundreds of metres away with no loss of quality. If the KVM is delivered on CAT5 it can be patched and so the system becomes flexible. If a computer in one studio needs to suddenly be available in another, a single CAT5 patch will facilitate this compared to traditional long KVM cables which are wired direct from place to place and would need replacing.

The CAT5 standard also allows for other modern IT devices to be installed in studio areas without the need for bespoke direct cabling. Maybe someone needs a laptop to be plugged into a network point, or a studio webcam to be installed.

Once the studio build has been completed it will inconvenient to run more cables and so the design and build stage is key to flexibility. If the station design is to use digital mixing desks rather than analogue, it may be the mixer itself requires IT connectivity and the CAT5 system will permit that.

When installing the CAT5 cabling make sure that it is tested to full CAT5 specifications and make sure there is plenty of space capacity for future expansion. A radio station built in 1995 will probably not have any CAT5 cabling as the standard was not commonly used. A station built in 2005 should have at least 24 CAT5 ports per studio.

CAT5 in Racks

A CTA will contain a number of racks, each of which may contain a number of computers requiring connectivity. There is a choice of cabling the CAT5 direct between racks, or alternatively to provide each rack with its own CAT5 patch panel, delivering 24 ports of CAT5 to each. Initially this feels like overkill, but the speed at which the CAT5 ports are used up is surprising.

If a single rack is to contain 5 computers, all of which are to be delivered to a studio area, then that will require 5 CAT5 patches to the studios. The machines themselves will need at least one CAT5 connection for network connectivity. Some machines such as servers may require more than one network point. In this case 10 ports of CAT5 are required. Allowing for expansion and changes in technology, 24 ports per rack is prudent.

CAT5 Patch Panels

As with Jack and Krone frames it will be necessary to have a CAT5 patch panel at some point. This will be the central hub where all CAT5 ports are presented and probably housed in one of the CTA racks. If the CTA is small it may be convenient to wire each machine or extender through the rack structure directly to the CAT5 panel. If, however, the installation will be large, and additional flexibility is required, it may be preferable to present a small CAT5 panel in each rack, which in turn is then presented at the master CAT5 patch frame.

This example shows a large installation where each rack requires many CAT5 connections and so has its own patch panel, which in turn links to the master frame.

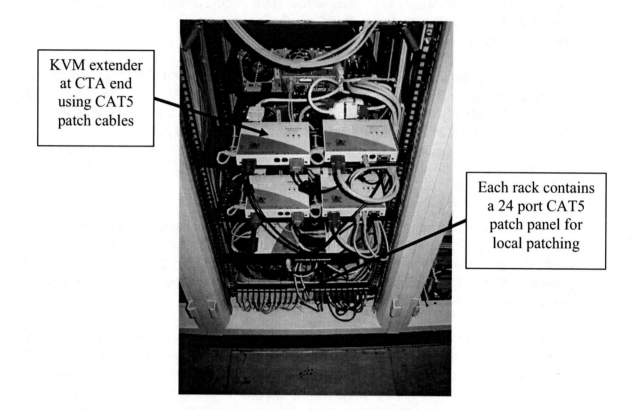

KVM extender at CTA end using CAT5 patch cables

Each rack contains a 24 port CAT5 patch panel for local patching

Figure 54 Example of structured CAT5 within a rack

As with Jack and Krone frames it is well worth keeping a master record of all the path routes and also of each individual patch.

In this example the route from the rack to the studio is described below.

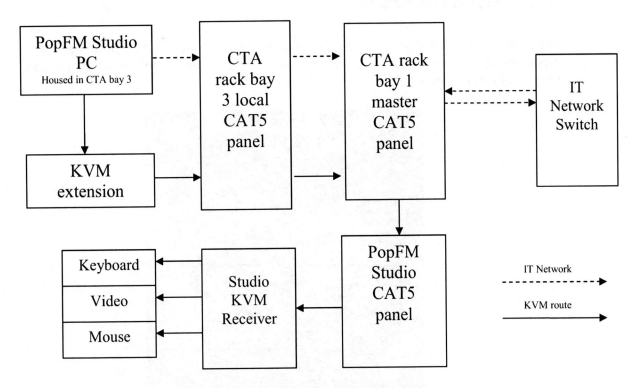

Figure 55 Detailed example of CAT 5 used for KVM and network connection

There are two routes required for this example. The IT network is required because the PopFM Studio PC needs to be on an IT network. The route is shown with the dashed line, a CAT5 patch from the machine in CTA bay 3 via the local bay 3 CAT5 panel. This panel links to the CTA bay 1 master CAT5 panel where it is patched in the Network Hub/Switch/Router housed locally in bay 1.

The solid line shows the CAT5 cable system used to patch the KVM of the studio machine to the studio. The local KVM cables connect within bay 3 to the KVM extender. The extender is patched locally to the CTA bay 3 CAT5 patch panel. The bay 3 panel is linked to the CTA bay 1 master CAT5 panel. The master panel has links to the studio panel, and a patch is made. The studio panel is then patched into the KVM receiver and in turn the individual Keyboard Video and Mouse services are presented.

This is a very common installation, and the flexibility is obvious. If a studio machine breaks a new one can be patched in either by swapping a patch lead from another studio machine or by simply removing it and installing a new one, all outside of the studio space, where time and access can be difficult.

If a new service or machine needs to be installed there is no need to install bespoke cables and systems as standard patchable systems are already in place, reducing running and installation costs.

CAT5 Patching Standards

When presented with a patch bay of several hundred cables, it can be confusing, especially if they are all the same colour. If some services are more critical than others then a colour code may help.

Figure 56 Suggested colour scheme for CAT5 usage

The colour scheme used here is

Red	Office IT Network
Purple	Broadcast IT Network
Green	Telephones/ISDN
Blue	Studio Mixer IT Network
Yellow	KVM
Grey	DC and RS232
Black	Audio

Not all stations will use as many colours as this, however clarity is the key and your colour scheme should be designed to help you.

Central Technical Area

The Central Technical Area (CTA) is the engine room of any modern radio station, it is where the clever stuff lives.

In most cases it will house, the electrical supply, UPS backups, the air-conditioning controls, the phone switch, the lines to the transmitter, the racks of studio equipment, the office servers; everything!

On a rough calculation based on a sample of 10 radio stations, it was worked out that each CTA needs 6U (U is a standard measurement of space taken up within a rack) of rack space per member of staff working for the station. Installations are all different, however, it follows that a small station with few staff members will need little in the way of resources and equipment and a more complex station with more staff will need more.

Rack Requirements

A medium sized station like PopFM may need 5 racks of equipment in its CTA and the layout could be:

Bay 1: Station master CAT5 patch panel, IT Network, Office phone switch
Bay 2: Computer Play-out servers and studio play out machines
Bay 3: Off air receivers, master clocks, ISDN equipment, audio processors
Bay 4: Studio talkback systems, Studio Phone in systems, studio amplifiers
Bay 5: Distribution amplifiers, monitoring, outside source switching, jack-field

The equipment list will be specific to the needs of the station and will change, however, the basic requirements are covered.

The starting point for the CTA is make sure there is enough space to house the racks you need and that you have space to expand. Once the CTA is full there is not much to do apart from knock it down or remove some equipment. Expansion space is critical.

Make sure there is space to work both in front and behind the racks, a minimum of 1 metre should be available to be able to manipulate equipment into and out of the bays.

A CTA can be dull and tedious, however, with a small amount of good design it can be made into an architectural feature. For example, glass doors opening into office or lounge spaces, solid glass walls, funky lighting and plenty of flashing lights all help. This can also be a good way to reduce the space taken up by the CTA on a floor plan as the 1 metre work space front and rear could also double as corridor or general space in the building. Consideration needs to be given to the health and safety impact of having a rack door opening into a corridor.

CTA Air Conditioning

The CTA needs to be very well air conditioned.

A studio may have 20 items in it generating heat; however a CTA will have 20 items per bay generating heat.

Detailed calculations need to be made by the cooling consultants. For a CTA of 5 racks 6KW of cooling should be about right. As most CTAs are sealed rooms an air conditioning failure will be catastrophic and so prudence would suggest that two separate cooling systems should be employed, one to act as a back up to the other. It may be that for a 6kW system, two 3kW units could be used. Should one fail the other would struggle to cool the area, however it would not be as catastrophic as a total failure.

The position of the cooling units should be considered. Wall mounted cassettes will take up valuable wall space which could be used for housing electrical supply panels or Krone Frames. If you have sufficient wall space this should not be a problem. Also consider that from time to time wall mounted units will leak water as condensation forms, positioning these above any electrical or electronic systems or cables could be problematic.

If wall space is at a premium, then a ceiling mounted unit may be in order, it should be positioned away from anything that may restrict access to it. Access must be available for servicing of all equipment.

Thought also needs to be given to the airflow in the room. The rack bays will create a chimney effect, funnelling heat directly upward. Equally a row of racks will create an artificial wall which may prevent air flow from one side of the room to the other. A system that can blow air to both sides of the racks from above may offer better cooling. Advice from the cooling consultant or contractor may be useful.

Air Conditioning on UPS supplies

If the main power to the building fails, and the CTA or studio air-conditioning is supplied by the UPS power system, then cooling should continue regardless. It will however be a significant drain on the UPS batteries and result in a shorter support time. On that basis considering the probability of a power problem, it might be reasonable to power the system from standard office power. If there is a power failure it may be short enough for the heat problem not to become a significant issue. If a split or dual air-conditioning system has been employed it could be half powered by office mains and half by UPS. This would result in some cooling and a lower drain on the life of the UPS batteries.

CTA Location

In the PopFM example the CTA is placed next to the studios

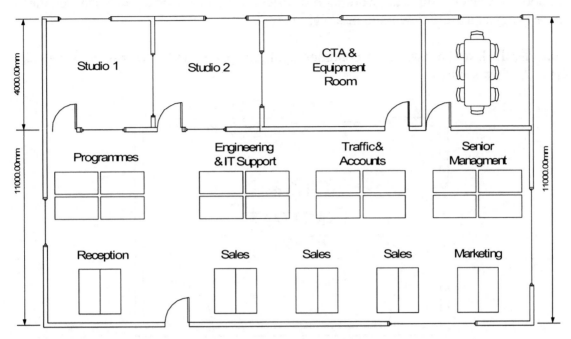

Figure 57 Position of CTA next to studios

It can in reality be placed anywhere, however there are many considerations. If the CTA is remote from the studio locations then it will be very difficult to see people working in the studios. This could be an issue during work on the live broadcast systems. Should an error occur resulting in a studio problem an instant visual cue from a studio worker will help minimize the time taken to understand and therefore fix the problem.

A CTA close to the studios also makes it much easier to run new cables and will reduce the cost of the overall cable during install. Installing more cable runs than you need during construction will reduce on-going costs should you need to expand.

If the CTA is several floors away from the studios then costs will mount, however, this is a good reason to install more cable as a future cable run will be very costly.

CTA Layout

A good CTA layout will make working life easier and make the difference between the space being a pleasure to work in or a nightmare.

Consider the layout of power, air-conditioning, krone panels and patch panels as well as the racks themselves.

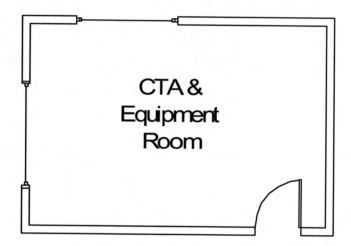

Figure 58 Empty CTA

The room is made up of several key things, two windows and a door. The windows offer daylight and a good line of sight into the studios, which is always useful when working on live systems that may effect on-air.

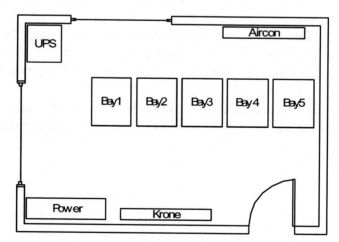

Figure 59 Populated CTA

This layout makes good use of the windows, allowing light into the room from outside and also uses the line of sight into the studio area. The air-conditioning is behind the bays blowing cold air directly into them for cooling. Power handling and krone systems have a good long wall to be mounted on allowing some space for expansion. The UPS which is noisy and bulky is housed alone in the far corner of the room.

All of these systems will to be interconnected and the space between them allows for tidy cable runs.

In an ideal plan, the power and audio systems should be kept as far from each other as possible. This has been achieved by locating them at opposite ends of the room and will result in lower mains electrical noise being induced into the audio systems.

To continue the separation of audio and power, the room could be designed to carry power in overhead cable trays and the audio and signal cables down at floor level. The room itself will be built on computer or raised flooring. This will allow the cables to be housed under the floor and out of the way. It also protects them from day to day damage.

Ideally the floor will be raised at least 200mm to allow enough space for the cable paths.

The racks bays need to be bolted to the floor as per the manufactures instructions, to prevent them from tipping or falling. They should also be bolted to each other creating a large solid structure. This will give them the strength to remain rigid and also add some additional safety. Each rack bay will need to be earthed and attention should be paid to avoid earth loops.

The Krone frame should be mounted to stand off the wall by about 100mm and to stand above the false floor by 300mm. This will allow the audio cables to enter the frame from below and have space behind the frame to be presented at the correct position, as well as making it possible to get access to the frame to work on it.

The power handling will need to be mounted directly to the wall, so the wall itself will need to be strong enough to take the considerable weight. There needs to be sufficient space above and below for the cables to access it. The cables will probably be armoured and need considerable space to bend and get into the distribution units.

The position of the racks in the middle of the room allows sufficient space to work front and back. The racks could be as much as a metre deep. An item of kit loaded into it could also be as deep as a metre. The working space front and rear needs to allow at least a metre, if not more, to physically raise the item to the position and slide it in.

When choosing racks for the CTA consider the amount of cable management there will need to be. Most rack mounted items are not a metre deep, however the additional space offered by extra deep racks means that careful and tidy cable management can take place. Poorly cabled racks are unpleasant to work in and increase the chance of a technical accident where a critical cable is knocked or unplugged resulting in the station going off air.

For ease of use it is traditional to nominate one side of the entire row of racks to be the front and one to be the rear. If you are displaying the equipment as a feature, then the front side would be the one that can be seen. It will contain the most flashing lights! Also the rear of the racks tends to look untidy, even when often they are not.

A single rack standing alone in an open environment, like an office, may benefit from front and rear doors, as well as side panels. In a CTA environment this tends to be unnecessary as the sides will hamper any cable routes between the racks (although they should only ever go below the false flooring) and front and rear doors offer little benefit verses cost. If you have racks on display or racks with complex cooling then they may be necessary. Un-needed rack doors always end up being stacked against a wall and it would simply save you money and space not to buy them.

Individual Rack Layout

A good well planned rack layout will prove easier to work on, and a poor layout will make life harder.

Early design plans should lay out where equipment will be housed. This allows the kit to be mounted quickly and easily. It will also provide a template for where cables should be presented. In many cases it is easier during installation to pre-wire the rack before the equipment is even mounted.

U number	Description
48	Mains Distribution
47	
46	Spare
45	Spare
44	Spare
43	
42	
41	
40	Phone Switch
39	
38	
37	
36	
35	
34	
33	
32	
31	Spare
30	Spare
29	Spare
28	Brush Access Panel
27	
26	
25	
24	
23	
22	CAT 5 Patch Panel
21	
20	
19	
18	
17	
16	
15	
14	Brush Access Panel
13	Spare
12	
11	Office IT Switch
10	
9	
8	
7	Internet Delivery
6	
5	Spare
4	Spare
3	
2	Phone Line Delivery
1	

Figure 60 Map of equipment in a rack

This shows an outline plan of the racks layout. Items that need to be worked on regularly such as the CAT 5 panel are placed at a reasonable working height, whereas items that will not be worked on regularly can be placed higher or lower

The mains distribution is in the top 2U of the rack. This is because once it is installed it will not be regularly worked on.

The phone switch is next in the top half of the rack, as once installed it is not going to be regularly physically worked on. Most modern phone switches tend to be remotely configurable via a windows interface, either locally via an interface cable or remotely from anywhere in the world.

The brush access panel below the phone switch will allow any cabling that needs to come from the rear of the rack to the front of the CAT5 in a controlled manor.

The CAT5 panel is placed half way down the rack as this system will need regular work. CAT5 patching should not be a daily occurrence but will be at least weekly.

There is space provided above and below the CAT5 panel to allow expansion, should more studios or equipment be added.

The office IT switch is also at an accessible point in the rack where it can easily be worked on with space above and below for any expansion that may be required. A brush panel is provided for easy cable access.

The phone switch CAT 5 panel and the office IT switch will produce a significant amount of patch cabling and thought at an early stage as to its cable management is important. This can come in a number of different forms, either entire 1U panels between the units, or cable management hoops that sit at the edge of each panel. Without this management the cable will look untidy and be very difficult to work on.

It may also be worth creating a CAT5 map, similar the one used for recording Krone records, to keep track of the cables and services provided on it.

The bottom of the rack is reserved for external connectivity. In the case of the phones and internet this could be a terminating unit from the Telecom provider which then patches into your phone or IT systems.

Creating an overall rack map, of kit and layouts will also be helpful when taking support calls out of hours. Having the information to hand will assist talking people through finding equipment they may not be familiar with.

The backs of the racks deserve as much attention as the front. This is where the real work takes place. To make the working environment as pleasant and reliable as possible good cable management must be employed.

The racks are designed to take cable trays up both sides of the rear of the rack. This will physically support the mass of cables in each rack.

Where possible power and signal cables should be kept separate, power up one side of the rack and signal on the other. This is not always possible. Power inlets on items of equipment never quite manage to all be on the same side, however, choose the most common side and stick to it. It is perfectly normal for mains to cross from the tray on one side only to be plugged in at the far side. The same is true for signal cables, be they audio or IT.

In some cases it is worth installing "tie-bars" which stretch from one side of the rack to the other, close to the cable tray. This will allow for cables to be supported as they cross the rack from the cable tray to the equipment.

Consider where the cables will enter the rack. If it is from the top then a huge item of kit at the top of the rack is going to get in the way, this may be better placed further down or at the bottom. Equally if cables enter from the bottom then care should be taken there too.

The picture on the right shows racks with mains cable entering from above and remaining separate from the signal cables entering from below. It also shows a dual mains system in use with a main and reserve power feed to each rack.

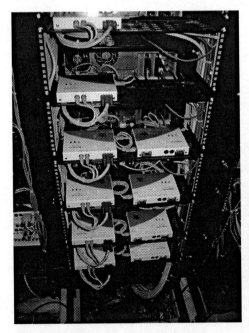

Figure 61 Cable access with a rack

The picture on the left shows the rear of a rack housing studio computers. The cables enter from the bottom of the rack and are immediately managed onto the right hand (from rear) cable tray. The racks themselves are deep to offer sufficient space to mount the KVM extenders on shelves to the rear of each machine. The left hand cable tray is used to coil and manage the local connection cables from the computers to the KVMs. Notice the rack to the left has an entire CAT5 patch panel just for the rack, which in turn feeds off to the master CAT5 panel.

Figure 62 Tidy rack on display

In comparison the front of this rack appears as a neat technical install and is on display, whilst all the cable management takes place behind.

Full CTA Layout

The full CTA layout plan will allow the design to be looked at in detail before the system is built. It will help check that there is enough space to install all of the equipment and also to see if there is any space available for future expansion.

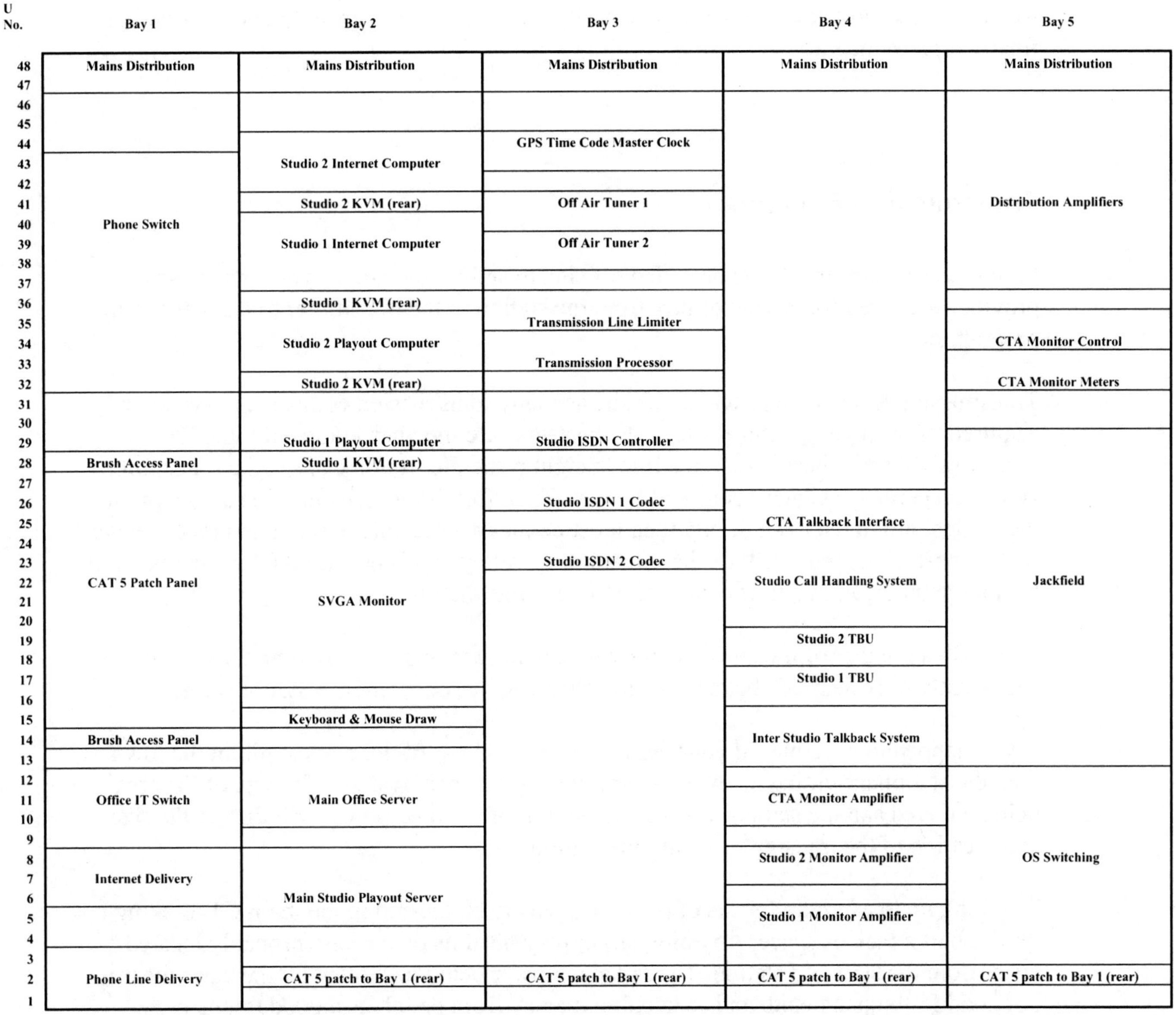

U No.	Bay 1	Bay 2	Bay 3	Bay 4	Bay 5
48	Mains Distribution	Mains Distribution	Mains Distribution	Mains Distribution	Mains Distribution
47					
46					
45					
44			GPS Time Code Master Clock		
43		Studio 2 Internet Computer			
42	Phone Switch				Distribution Amplifiers
41		Studio 2 KVM (rear)	Off Air Tuner 1		
40					
39		Studio 1 Internet Computer	Off Air Tuner 2		
38					
37					
36		Studio 1 KVM (rear)			
35			Transmission Line Limiter		
34		Studio 2 Playout Computer			CTA Monitor Control
33			Transmission Processor		
32		Studio 2 KVM (rear)			CTA Monitor Meters
31					
30					
29		Studio 1 Playout Computer	Studio ISDN Controller		
28	Brush Access Panel	Studio 1 KVM (rear)			
27					
26			Studio ISDN 1 Codec		
25				CTA Talkback Interface	
24					
23	CAT 5 Patch Panel		Studio ISDN 2 Codec		Jackfield
22		SVGA Monitor		Studio Call Handling System	
21					
20					
19				Studio 2 TBU	
18					
17				Studio 1 TBU	
16					
15		Keyboard & Mouse Draw			
14	Brush Access Panel			Inter Studio Talkback System	
13					
12					
11	Office IT Switch	Main Office Server		CTA Monitor Amplifier	
10					
9					
8				Studio 2 Monitor Amplifier	OS Switching
7	Internet Delivery				
6		Main Studio Playout Server			
5				Studio 1 Monitor Amplifier	
4					
3					
2	Phone Line Delivery	CAT 5 patch to Bay 1 (rear)	CAT 5 patch to Bay 1 (rear)	CAT 5 patch to Bay 1 (rear)	CAT 5 patch to Bay 1 (rear)
1					

Figure 63 Map of CTA rack equipment

At the design stage the layout plan also allows a simple asset check to make sure enough equipment has been ordered.

It is often difficult to decide whether to populate the racks with all of the equipment and then cable to it, or to install the cabling first. To date there is no right answer, it can help to have the kit in place so that cables can be loomed and made perfectly to length. Leaving a service loop on the cables can also be a good idea to make sure if the equipment does need to move up or down a little then it can, or even to allow the kit to be removed all the way out the rack whilst still connected. The reality of the situation is somewhere between the two, often placing the kit in the racks will help visualise the work that needs to take place and to check the expected connectors and length of cable. Remove the kit to allow the racks to be easily worked on. Then install the cables and then finally replace the kit.

Transmission Equipment

In most cases transmission is provided off site, meaning that you or your transmission provider will need some kind of link from the station to the transmitter to carry the radio programme.

Transmission tends to be provided off site as many transmission companies have already acquired the best geographical sites. The best sites are the ones at tops of hills. FM transmission works best based on a line of sight principle. If you can see the transmitter you can receive the signal being sent from it. Therefore if the transmitter is at the top of the biggest hill in the town or city then most people will be able to see it and its coverage will be best. If the transmitter is based in a deep valley and is obscured from the main city by a large mountain, then its coverage will be somewhat limited.

There are a number of transmission providers who specialise is providing these services, and usually provide much better coverage than can be achieved by a station alone.

If you happen to have placed your radio station at the top of the biggest hill in the town then the station providing its own transmission is very achievable. The size of the area being covered and the stations maximum amount of radiated power will dictate the size and location of the transmitter within the station.

Anything up to a hundred watts of power can easily be located in the main CTA taking up maybe half a rack of space. Anything larger may need its own room, probably located away from the staff and station. The radiated power being generated by the equipment is not good for human health and so keeping it away from people is a good thing.

The cable that runs from the transmitter to the antenna will need to be a short as possible and will be large in diameter. This is important as large diameter RF feeder cables do not easily bend round corners and so the cable runs need to be carefully thought out. They should be kept away from people and electronic equipment. The cables over the distance of the run will lose power which will interfere with electronics and people.

For this reason it is not uncommon for the transmission equipment to be located in an out-house at the base of the transmission mast.

If the equipment is housed in the CTA it needs to be low power, treated with some care and kept as far away as possible from other electronic systems. It may need to stand in a rack on its own, which is fully shielded with solid metal doors, sides, tops and floors that will prevent leakage of Radio Frequency radiation.

CTA Equipment List

For completeness a full CTA equipment list should be compiled to assist in the purchasing process. A large purchase will always attract a discount compared to several small purchases.

The list can also be broken down to assist the reader and ensure clarity.

Mains Distribution, in each rack to provide power from a single (or dual) feed to many items within the bay, each output will be individually fused.

Phone Switch, to allow office and studio users access to telephone lines and services.

CAT5 patch panel, where all data services will be cross connected, between racks, desks and studios.

Office IT Switch, Connects all computers together to provide a computer network.

Internet Delivery, where the ISP provides an internet circuit for your use.

Phone Line Delivery, where the telephone company delivers your phone lines for the radio station, maybe DEL, POTS, ISDN or Q931 circuits.

Studio Playout Computer, the machine that allows the presenter or producer in the studio to play songs, jingles and commercials off a computer system.

Studio Internet Computer, whilst the playout computer plays the songs, the presenter may wish to get email or check news stories on the internet whilst preparing for the next link.

KVM, extends the computers keyboard, video and mouse from the CTA to the studio, allows the studio operation to be silent, compared to fan the noise generated by a PC.

SVGA monitor and keyboard, to allow monitoring of servers and machines in the CTA. It may work in conjunction with a KVM switcher to allow access to several servers or machines in the CTA without having to move cables round.

Office Server, a mass storage device to hold all office files and folders upon, will also perform administrative role, such as providing security to those files and requiring people to log in and out.

Playout Server, holds all of the songs and databases that the station plays. The studio playout machines get all their data from this server.

GPS time code master clock, receives time data from the GPS network. It acts as master passing that time code to IT networks and studio clocks to always ensure the time is correct.

Off Air Tuners, allows the station to receive its own and other stations transmissions, a good check to ensure station is on air.

Transmission Line Limiter, ensures the signal being passed to the transmission lines doe not exceed maximum tolerance, which would cause distortion to be introduced to the programme.

Transmission processor, can often be found at the transmitter rather than the station. It provides a dynamic EQ curve to the general sound of the station. As it is dynamic, it will not add more base when it detects there is plenty and will add more treble if it detects there is not enough. It will also set basic sound levels and correct then with an Automatic Gain Controller (AGC). When the programme is too quiet it will make it louder and when the programme is too loud will make it quieter.

Studio ISDN Controller, allows different studios to control the centralised (based in the CTA) ISDN Codecs.

Studio ISDN Codecs, a coder and decoder (codec) which will convert and compress audio into a digitised stream to be passed down an ISDN line. At the other end a codec will decompress and convert this stream back into an audio feed. This is a standard method of connecting remote studios together to pass audio between them, and also a very good method of providing an Outside Broadcast link from the location back to the studio for transmission. These units require an ISDN line from a telecoms provider to function.

Talkback, a method of providing voice based communications from one studio to another, or from a studio to the CTA, without the signal or conversation going on air.

Studio call handling system, allows the station to hold a phone in, and will pass the one selected phone line from the many delivered to the station to the TBU.

Studio TBU, converts phone calls from the telephone company standard to an audio level standard. It will also offer significant improvements on normal phone quality by balancing the amount of signal sent with that it receives, and works very well when supplied with a clean feed from the studio.

Studio Monitor Amp, provides the amplification required to boost the line level studio audio to a level that will power loud speakers, often housed outside of the studio to prevent presenters turning it up too much.

Distribution Amplifiers, takes one source signal, such as the Off Air Tuner and makes it available to many destinations, such as studio 1, studio 2, CTA, office etc. Also offers some protection by isolating each output. In the event that one fails due to a problem the rest should continue working.

CTA Monitor Control, a switching box to allow important audio sources and feeds to be monitored in the CTA, similar to an outside source switcher. This allows technical checks of feeds such as studio, transmission and reception to be made to assure quality and also to help diagnose problems.

CTA Monitor Meters, a set of meters measuring the signals levels as selected by the CTA Monitor Control.

Jack Field, a flexible audio patching solution to allow temporary changes to the designed audio routes and paths created at design, also useful for monitoring and resolving problems.

OS Switching, routers that select an audio source on demand using a controller to allow an Outside Source channel to be switched between many sources in a studio.

Commissioning and Testing

Physically building and putting all of the equipment together is a significant part of building any radio station. The success of the project will be based on if the equipment works in the expected way. If the station looks fantastic but simply doesn't work then the project will have failed.

To have spent several months working on something so complex only to find it not working is deeply disappointing and so testing each part of the installation as it is constructed is a sensible way to proceed.

Structural Installation

As the structure of the station is being constructed, probably by external contractors, keep a detailed check-list of all the works and of all the key components per area. For example, break the work down into areas and features.

Studio: Floor, Walls, Ceiling, Fabric Finish, Doors, Windows, Air-Conditioning, Sound-proofing.

As the contractors are working on these areas check their work, be familiar with what they are meant to be doing and what they are actually doing and finally what they have done. Raise issues with the site foreman as they arise, agree on a resolution and a time scale and make sure they are completed. In all cases the contract and specifications should be consulted to make sure the work meets the expectation of what is being paid for.

It is much easier to resolve any problems at an early stage, rather than weeks later when the building project has actually turned into a working studio and all of the builders have left site.

Go round and check that the floor has been correctly finished, there are no holes, it doesn't creak or move. Make sure that the fabric finished walls are smart and well presented, not loose or creased.

Doors should close easily, of their own accord first time and without help. They should seal to create an air tight gap, a traditional test of this, is to see if a piece of paper can be slid between any gaps or joins. If it can, there is clearly an air gap.

The sound-proofing of a room is harder to test, however, it is critical and should be checked regularly through the build. The original contract should specify how much sound isolation is to be achieved. Simply standing in the room and listening to external noises will give an instant feel for success.

Specific tests can be carried out by acoustic engineers to measure the precise level of isolation achieved, although in many cases this can be done at much lower cost by the building contractors or station engineers.

Take a Sound Pressure Level (SPL) meter and a noise source, most probably a PA system. In the studio run the PA system at a loud volume and measure the level produced at least a metre away on the SPL, set the system to produce levels around 80 to 90 dBA.

Stand outside the studio door and measure the noise levels with a SPL.

The measured Studio Level minus the measured External Level equals the amount of isolation achieved.

If you are aiming for 30dB of isolation and the noise source is set to run at 80dB, then the measured area outside of the studio should record no more than 50dB.

For completeness this should be carried out in many different locations, outside the studio doors, in the next door studio, and basically any space that borders the studio under test. This is far from a definitive test, but it is a good rule of thumb.

More detailed tests can be carried out at number of frequencies with sine wave test tones. This will test the isolation at different frequencies to find out how well the studio performs across a frequency range. It also may help to spot any difficult resonant points that may be resolved at an early stage.

A basic set of frequencies might be, 40, 80, 100, 200, 400, 800, 1k, 1.5k, 2k, 4k, 8k, 10k, 12k and 15kHz. A graph can then be produced for each area tested.

If the isolation measurement is critical then a specialist company should be employed to provide this measurement.

At an early stage in the build it will be reasonably easy to add extra sheets of plaster board or block up more holes with silicon sealant or plaster, however, once the final finish fabric is being added this will become complex and costly.

Electrical Testing

Basic testing should be carried out as each cable is installed. Cable identification and a resistance test should prove the right cables are going to the right places. If the testing takes place before the fabric of the studio is completed then any problems should be easy to resolve. Once the fabric has been installed then the job of removing and fixing a cable will be difficult and costly. A detailed schedule of all cables and tests should be kept. The electrical installation contractors should provide a final installation safety test certificate, which should be filed for future inspection.

Electronic Testing

This is the area where the broadcast engineer should concentrate the most, as it is the area he will be working on. When installing a studio it is very likely that hundreds of cables will be presented and terminated, which in turn will mean many thousands of pairs of cables.

A detailed schedule of cables run, from place to place, should be made before any cable is installed, and will form the basic installation map and test procedure. Each cable should be labelled at each end. As each cable is made off onto its termination connectors (be they Krone or XLR or D-type) they should all be tested with a basic resistance loop test. This will identify each cable and pair, and will indicate any cable breaks, cable swaps or shorts. Once this is done, there will be confidence in the cabling so that any further problems will be down to local patching or equipment faults.

In a station where there may be five or ten thousand individual cable pairs in the system, a fault half way through the installation may takes days to find. Test all cables as they are installed and keep records.

Broadcast Systems Testing

Once the installation is complete it is should be tested as a system. These tests should consist of phase and level measurements for the system components. If the installed system is AES digital then specialist equipment is required for comprehensive testing, however, basic testing can still take place. If the system is analogue then thorough testing can take place without specialist test kit, although for analogue it is often more critical for good test kit to be available.

The purpose of the tests is to prove that all the major system components are functioning and that the links between them are good. The test should prove that known audio levels, at several frequencies and phase, are faithfully reproduced from an input such as a CD player to an output such as the studio outputs.

Start by listing all of the known inputs and outputs within the system, and then devise a test procedure and checklist to methodically go through all these items.

In PopFM Studio 1, the original kit list can be the starting point. It may also be useful to use the Jack Field or Krone Field as the major test points.

This diagram will help producing the test list and also help find good test points.

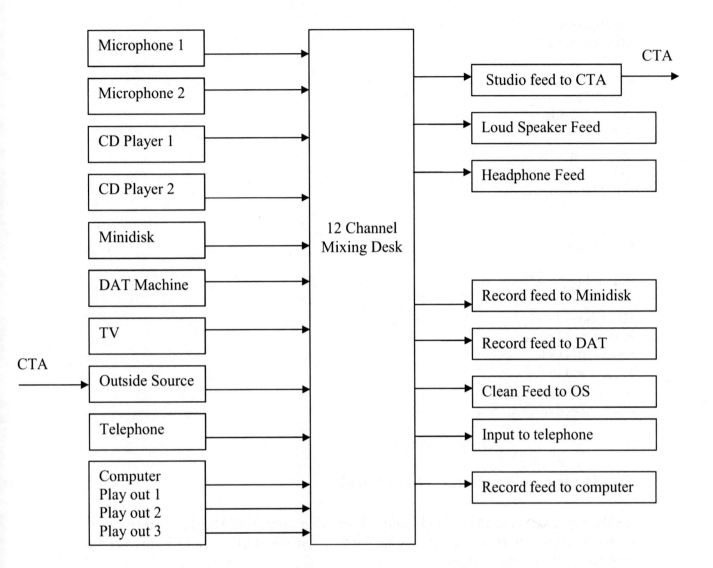

Figure 64 Studio system diagram

Sources:

Microphone 1
Microphone 2
CD 1
CD 2
Minidisk
DAT Machine
TV
Outside Source
Telephone
Computer Playout

Destinations:

Studio Feed to CTA
Loud Speaker Feed
Headphone Feed
Record to Minidisk
Record to DAT
Clean Feed to OS
Input to telephone
Record Feed to Computer

Studio Line-up Procedure, Line Level

Set the mixer inputs to all be at 0dB gain, which principally should be the fader (for that source under test) to fully open (assuming the faders stop at 0dB and not +10dB) and the gain set to 0dB on each channel.

The first line level source is CD1. Take a known standard or pre-produced test CD, and play it at the nominal 0dB (PPM4). Depending on the station wide level standards you are using this may be -22dBFS on the CD player.

Insert the test measuring set into the XLR outputs of the CD player, this will give the true audio level out. Then calibrate the CD outputs (available on most professional machines) until the audio level measures 0dB (PPM4). Also ensure that left and right are the correct way round. Standard test CDs will have left-right identification tracks. Check the CD players output is in phase. Plug the CD layer back into the studio cabling and then move to the Jackfield listen points for the CD player and repeat the test. Any mismatch in level would suggest a cabling problem.

Set the mixer channel level for the CD player to be 0dB gain on the fader and 0dB on the gain pot, ensure there is no left right pan.

This should then register on the mixer meters at PPM4 or Nominal 0dB. If the reading is not 0dB then the audio channel on the mixer will need to have its trim settings calibrated. When the mixer reads 0dB, left, right and phase are correct, the mixer outputs themselves can be checked.

Plug the test kit into the main mixer output at the Jackfield and check the levels read 0dB, left and right are the correct way round and the audio is in phase. If it is not check the cabling from the mixer to the Jackfield, if that is good it may be necessary to calibrate the main mixer outputs. Repeat the process for all of the mixer outputs including, clean feed and record outputs.

Lining up analogue audio channels can be difficult and a tolerance of 0.5dB is acceptable on all channels, although better should be aimed for.

Once one channel has been calibrated, repeat the sequence for the rest of the input channels. When there is confidence in the outputs of the mixer the rest of the channels can be calibrated at the input stage alone, with occasional output level tests for confidence.

Continue the process until all inputs and outputs have been tested according to the test documents sources and destinations list, noting the levels at each test. Once the process is complete the mixer is calibrated.

It is important to use good test equipment as these levels will now become the station wide settings. Ideally a calibrated test kit should be used as this will align the station to known world wide standards which will help when connecting to external sources and destinations such as transmitter feeds.

Studio Line Up Procedure – Microphone Level

Microphones can be complicated as there is an additional gain stage to lift the microphone level up to line level. This setting will need to be configured by hand to find the best gain setting for the microphones being used. Some microphones may need as much as 60dB gain, whereas others may only need as little as 30dB.

Tests can be done by removing the microphone and supplying test set signals at -30dB (or whatever setting is required) and this is useful for proving phase and recording the results.

If an external mic amp is used, it should be tested with mic level into it to obtain a 0dB output and then checked on the mixer to ensure consistent results. If a mic processor is used it will affect the audio levels by design and should be tested in bypass.

CTA Line-up Procedure

The line up procedure for the CTA is little different to that of the studio, a case of listing all sources and destinations, passing known level audio through them and measuring the results. The measurements from input to output should match and where they don't they will need calibrating or investigating to find a fault until they do.

The CTA will house more equipment and so the test procedure will be longer than that of the studio, and again detailed check lists should be composed to ensure nothing is missed.

At the very least the basic broadcast train should be tested.

Sources:
Studio 1
Studio 2
Studio Switcher Output

Destinations:
Studio Switcher Input
Transmitter

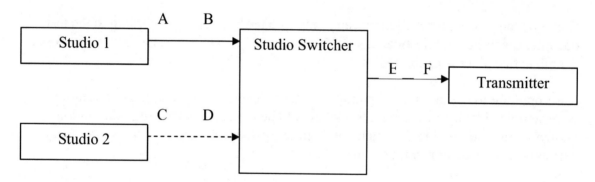

Figure 65 Basic broadcast chain to be tested

With a known level source of 0dB (PPM4) in Studio 1, check the level in the CTA on the Jackfield at the Studio 1 output point A. If the cabling and Distribution Amplifiers are all operating correctly this should read the same as the input of 0dB. Also check out for Left and Right and Phase.

Repeat the test for Studio 2 outputs at point C. Then test the inputs to the Studio Switcher at points B&D, then the outputs at point E. Finally the audio levels being presented to the transmitter at point F. When the full chain is under test from Studio 1 point A to transmitter point F, the levels presented to the transmitter should be within plus minus 0.5dB of the level generated in the studio, with Left, Right and phase still being correct.

At this stage it is known the system will faithfully reproduce on-air the source generated in the studio.

Transmission Testing

Most stations have a period of test transmission before they go on air. This usually means test broadcasting before programmes proper start. This will be one of the only chances to perform a technical test on the transmission system.

Level, left, right and phase tests should be carried out. If a transmission processor is part of the system it should be tested, however, it will affect levels and may also adjust left right balance and phase as part of its design. During first stage testing it should be placed into a bypass mode, or even taken out of the chain.

It is also common for the government regulator to attend early testing to ensure that the system performs within specification and that it does not create any spurious transmissions that breach the terms of the broadcast licence.

Working with the regulator is very important as they have the power to shut down your station. They may be tough and demanding, however co-operation rather than conflict will result in a faster and less painful experience. The government issued broadcast licence should contain details of how to contact the regulator and arrange testing.

Transmission

Transmission basics are simple, however the methods of achieving it rarely are. Radio Frequency (RF) broadcasts are often referred to as a black art as there are simply so many variables to easily predict how it will work and how many people and places it will cover.

Type of Transmission

There are many ways to broadcast these days ranging from good old AM to Satellite or the Internet. They all behave differently in terms of cost, performance and coverage.

Longwave

Longwave is a very low frequency Amplitude Modulation (AM) method of transmission that requires very high power transmitters of thousands of watts. The signal is of low audio bandwidth around 3 kHz, and is very noisy. The signal quality varies dependent on the time of day due to changes in the atmosphere dependent on the time of day. The coverage can be many thousands of miles from the transmitter. The running costs are vast as the power consumed is so high. Setting up a new Longwave transmitter is also expensive as the transmitting antennas tend to be multiple pylons erected over many hundreds of metres. Longwave tends to be the remit of national broadcasters wishing to reach an international audience

This is not a common transmission method for new radio stations, does not suit music and is only passable on voice.

Shortwave

Shortwave is a low frequency broadcast using AM modulation that requires high power transmission. The signal quality is dependent on atmospheric conditions and times of day. Coverage can be many thousands of miles depending on the power of the transmitter. Transmitter setups can be expensive as the antenna systems required are large and consume large amounts of power. The signal carries a low audio bandwidth programme, suitable for speech and not suited to music. Shortwave tends to be the remit of national broadcasters reaching an international audience.

Low Power AM

Low Power AM (LPAM) is a scaled down version of the common domestic AM services. Set up primarily for student and community radio services. LPAM is dependent on local geography. Power outputs tend to be limited to one or two watts of radiated power. In a flat geographical environment it can cover maybe 5 to 10 miles. In more built up areas this may be reduced to just one or two miles. Installation is reasonably cost effective and can be local to the radio station as the antennas tend to be a single mast similar to a lamp-post with a wound coil at the top. Audio quality is around 4 kHz. This is good for speech and passable for music, although it could not be described as good.

Coverage is dependent on atmospheric conditions and will vary between night and day. LPAM has been known to travel many hundreds of miles as it bounces off the atmosphere, however, was never designed to do so. Running costs are low as the power outputs are low.

Induction Loop

Induction loop was designed for very short range reception, often reaching no more than 10 or 20 metres from the antenna. It is often used by student or community radio stations to cover a residential hall or campus. It can be expensive to install due to the fact a cable has to be installed in a loop round the entire area to be covered. It operates in the domestic AM bands, and provides 4 kHz audio bandwidth, which is good for speech and passable for music. Induction loop suffers from local interference and can also induce the radio programme into other devices such as telephones. Running costs are reasonably low as power consumption is also low.

Amplitude Modulation

Amplitude Modulation (AM) was one of the earliest radio technologies after Morse code and spark gap generators. It has been with us for many decades and is still home to many services. Audio bandwidth is 4 kHz which is good for speech but poor for music, although many music services do operate on AM. Transmission systems are expensive as the kit consumes high amounts of power and therefore has to be large and complex. The antenna systems are similar to those of long-wave, although smaller still consists of multiple pylons spread across hundreds of metres of fields. The range can be around 100 miles, although fifty would be more common. Running costs are high due to the huge amounts of power consumed. Reception is effected by atmospheric conditions and is traditionally terrible at night.

Frequency Modulation

Frequency Modulation (FM) is the most popular high quality transmission method and has been around for over thirty years. It is ideally suited to speech and music, and can operate in mono and stereo. FM has high quality audio bandwidth of 15 kHz stereo and is stable and predictable in most atmospheric conditions. It works on a line of sight theory, which means if a line can be drawn between the transmitter and receiver, reception is possible. To get the best results the transmitter antennas are normally located as high as possible, on top the largest hill in the area. FM will reasonably penetrate obstructions in the line of sight, such as buildings, however if local geography such as hills obstruct the line of sight the signal will be lost.

FM can also carry data services such as RDS (Radio Data Service) that is often used to provide station name information to the receiver which can be displayed, as well as news headlines or other dynamic text messages on a small screen on the front of the radio.

Coverage of a reasonably flat area can be around 50 miles depending on the transmitter power. A small transmitter of 10 watts in a flat area could cover about 5 miles, a larger transmitter of 100 watts could cover about 15 miles and a large transmitter of 4,000 watts could cover about 50 miles. Power required to cover the target area is dependent on the height of the transmitter and the local geography. In a mountainous area, a transmitter of 4,000 watts may cover 10 miles in one direction and as little as 2 miles in the other. Any area not able to get line of sight, such as a valley on the other side of a mountain will have little on no reception.

Imagine replacing the transmitter with a light bulb, if the light can be seen, radio reception will be achieved, if the light cannot be seen because it is obscured behind a geographical feature, reception will not be achieved. If the transmitter is placed on top of the largest mountain, that will give the best chance of reception.

Digital Audio Broadcasting

DAB, Digital Audio Broadcasting, is one of the newest transmission methods. DAB transmission is provided by a Multiplex, which can broadcast several stations at the same time inside a large digital signal. Once the main digital signal is received many stations are then made available. DAB provides varying rates of audio quality chosen (normally on a price basis) by the station broadcasting.

A station broadcasting mostly speech may run at as little as 64kbs, a music station could run at 128kbs, and a high quality music station at 192 or 256kbs.

DAB also allows data services to be included within the transmission, providing for Dynamic Label Services (DLS), similar to RDS, as well as Electronic Programme Guides (EGP), and other enhanced data services. The DAB multiplex allows for a sufficient data rate that software can be delivered to users, to update tuners, talk to PC's or PDA's to provide additional data for other software applications.

DAB is designed to operate on a single frequency network. This allows multiple transmitters on the same frequency to provide additional fill in transmitters in areas either out of the primary range, or to operate as a large network.

FM was originally designed for a fixed receiver location, whereas DAB has been designed for mobile listening. The big advantage of DAB over that of FM is the lack of hiss or crackle. The service works or it doesn't.

Multiplex are complex and expensive and not usually affordable for a single station to set up. It may be a problem to get access on an existing multiplex or to find partners to create a new one.

Satellite

There are many satellite broadcasters, some specifically delivering radio to dedicated radio receivers. Some provide radio channels on what is primarily a TV service. Most tend to be digital and the quality can be very good. Delivery on TV platforms is an easy way for many to listen at home, but generally not easy when mobile. Satellite is a good extra service to offer on top of the primary delivery of AM, FM or DAB.

Digital Satellite also allows data services such as EPG to be broadcast along with the programme service

Cable TV

Cable TV companies are also delivering radio services similar to the Satellite broadcasters, again a good fill in option, but not available for mobile listeners. Cable TV also allows EPG and other data services to be available.

Digital Video Broadcasting

Digital Video Broadcasting companies are delivering radio services similar to the Satellite & DAB broadcasters. A good fill in option and can be available for mobile listeners but is not designed for it. DVB also allows EPG and other data services to be available.

Internet

Many stations are now making programmes available to the internet, and with the increasing availability of broadband. This allows good quality audio to be delivered. Some stations only broadcast on the internet. In some instances this is unregulated, meaning that fewer licensing issues have to be addressed. With the introduction of wireless internet mobile reception is becoming more of a possibility. Data services can be provided via the internet which can include additional audio streams, pictures of video.

Third Generation Mobile

Third Generation mobile phone technology is essentially delivering mobile internet, and therefore making internet radio available in more places. The cost of using 3G tends to be the only restriction of this point. 3G is a leading edge delivery platform and one not yet fully explored. Ultimately the services available could include pictures with audio on demand. The video of the song being played right now could also be included in the broadcast. Another option could be a buy now service where the song playing now can be purchased as a download.

With 3G any additional data service that can be imagined can be delivered such as EPG, downloads or extended data services to name but a few. Internet in the sky is what 3G is promising and there will be a price to pay, however, the connectivity is there and if you can imagine it, it can be done.

Transmission Processing

Every great radio station has a transmission processor. The aim is to make the station audio sound better. That maybe to ensure there is a consistent audio level or maybe a station sound style applied to every item of audio ever broadcast. In some competitive markets transmission processing is all that makes the difference.

If two stations are playing the same song, what will make the listener choose your station over the next? The answer is simply station sound. Many stations pride themselves on how their station sounds, and that is what a new station needs to compete with.

The role of the station engineer is to facilitate that sound. Never assume that as station engineer the problem is the engineers, the reality is that the station management will have strong opinions. The best way to deal with this is to allow the management team to have the sound they want.

Processing Folklore

Will the station sound make a significant difference to the ratings? If the station has a good sound compared to a competitor then it will not make the ratings difference. If it has a poor sound it may. On-air content is the key factor for ratings, however, processing will assist in the following areas:

A transmission processor is often just one unit, but could consist of many separate units set up in the following chain.

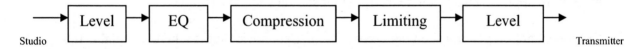

Figure 66 Transmission processing system diagram

Level

Each good station should apply an automatic gain controller (AGC) before transmission. This will correct any level problems where the source audio is too loud or too quiet. The transmitter can only work with the audio it is given. If the level is wrong there the listener will end up constantly turning the radio volume up and down. This will ultimately cause the listener to move to another station that has more consistent level. The AGC stage in the processing will automatically adjust audio levels to create a consistent sound.

EQ

Part of the station sound is the EQ applied to the audio. If the station is targeted at a young urban music market, then increased base may be required. If it is a speech service then it may be more appropriate to focus on the low to mid audio ranges to add power and dominance to the voice. A classical music station may not wish to change the EQ of the music and allow the natural sound to dominate. The processor will allow the EQ to be set across the whole audio feed.

Compression and Expansion

Compression is a method of limiting audio feeds that are too high, but is used conceptually the other way round as a method of boosting an audio feed. By boosting the level into a compressor and then limiting any audio over a certain level. The result will be a more consistent and full sound. Quiet signals are boosted at the input to the compressor and pass through the device louder than when they started. Levels that are too loud, due to the gain applied, will be limited back down to the correct levels.

Compression could initially be compared to the AGC stage, except the AGC tends to be slow to make up gain, whereas the compression stage is much faster.

Compression in the transmission processor is often multi-band, which means that the base, mid and treble can all be processed differently. This can be used like the EQ stage, where you may apply high compression to the base and treble and little to the mid audio band. The advantage with compression over EQ, is that it is dynamic to audio source, so where a signal has plenty of base little more will be added, but a signal lacking base will have more applied. This is where a consistent station sound is achieved.

Expanders are often confused with compressors. Expanders detect where a signal is too quiet and the processor will reduce the gain level over time, until ultimately it is turned off (gated). This is useful in speech where the presenter may pause between words. Rather than trying to boost the silence coming from the microphone the processor knows to reduce the level to compensate for any back ground studio noise that is really not intended for broadcast. Again expanders can be multi-band and result in separate gain reductions for base, mid and treble.

Limiting

Once all of the main EQ and compression has been applied a limiting stage will ensure that any unexpected level peaks are controlled.

Level Adjustment

The final stage is a level adjustment. Once the audio has been processed and limited, to ensure it stays within known level criteria, there is often a final level adjustment to be made to match the processed audio with the transmission equipment.

Transmission Systems

A basic transmission system will consist of audio from the studio source, processing and the transmitter.

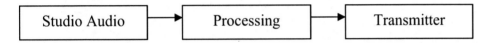

Figure 67 Basic transmission system

A more complex system will offer more options, including back up transmission lines from the studio in case of failure, as well as back-up CD players at the transmitter in case the studio has problems generating audio. There may also be back up processing and transmitters all set to automatically switch over in the event of a problem.

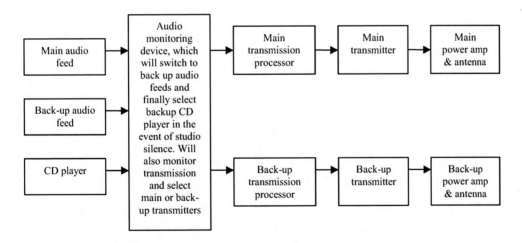

Figure 68 Advanced transmission system

Monitoring of transmission

An off air receiver in the studio or station is the primary method of checking the station is on air, however the transmission system itself should be designed to provide telemetry and alarms when it has problems. These may be passed to the transmission provider who will deal with any problems, and alert the station or engineering team of them. If the transmission is being handled by the station itself then local alarms should be generated that will report to the on air studio, so that problems can be passed to the engineer.

Glossary of terms

Much of any trade or business is made more complex by the industry jargon used, and radio is no different. This glossary of terms will help clarify some of the abbreviations, acronyms and phrases.

¼ Inch Jack	Unbalanced 3 pole audio connector for left, right and ground
13 Amp Socket	Name for a standard household mains connector at 220volts
3.5mm Jack	Unbalanced 3 pole audio connector for left, right and ground
Adderlink	Brand of KVM extender
ADPCM	Adaptive Delta Pulse Code Modulation, digital compression system used to reduce the size of audio files, can result in loss of quality if used at low bit-rates
ADSL	Asymmetric Digital Subscriber Line, internet connection, usual in the domestic market. Can be found in the professional market for low cost internet connectivity
Aerial	Device to allow broadcast or reception of radio signals
AES i	Audio Engineering Society, set standards and work to develop new audio technologies
AES ii	Common name for Digital Audio standard, derived from standards set by the Audio Engineering Society
AGC	Automatic Gain Control, automatically increases or reduces gain according to user set parameters to create a consistent audio level at the output
Air Conditioning	Cools or warms the air within a confined space such as a studio or office to create a consistent temperature as defined by the user
Alarm	An audible or visual indication of a critical problem with an item of equipment or a system
Alert	An audible or visual indication of a minor problem with an item of equipment or a system
AM	Amplitude Modulation, method of mono low quality audio transmission, with audio bandwidth of around 4 kHz
Amplifier (Amp)	Device to replicate an audio signal at a higher volume, usually capable of powering loud speakers
APF	Automatic Pre-Fade, when a source is Pre-Faded, the normal monitoring is automatically changed to follow the source being Pre-Faded
APTX	A digital compression standard usually found in studio ISDN equipment to reduce the data rate required to pass audio from one site to another. Common for low quality usage, but can run at high quality, although requires much more bandwidth

Arc Master	A system for remote controlling, dialing and connecting ISDN calls using centralized equipment
ATXM	A method of providing ISDN audio feeds for a computer user which will connect to studio based ISDN equipment
A-Type Jack	Unbalanced 3 pole audio connector for left, right and ground, most commonly used in domestic equipment
Automatic Pre Fade	See APF
Automation	A method of generating radio programmes unattended. Often refers to a computer playout system programmed to generate music or speech content.
BBC	British Broadcasting Corporation
Black Clock	Method of providing AES Digital Audio timing signals by feeding a silent Digital Audio signal which also contains time reference to a device
BNC	Bayonet connector for high frequency signals, such as radio frequency
Breaker	Electrical safety device to shut down power supply in the event of a fault, such as excessive current being drawn
B-Type Jack	3 pole balanced audio connector, provides ground, phase signal and anti-phase
Buss	A common signal path or point, where multiple devices can access audio signals or logical control signals. Often found within mixers, such as programme buss, or record buss.
Cassette	Cartridge or box containing reel of audio tape
CAT 5	Category 5 data connectivity system, used for interconnecting data, computers and telecoms systems
CD	Compact Disc
Clean Feed	A specific audio mix, generated when an outside source is used within a studio, a Clean Feed is returned to the external contributor which contains the full programme mix without the contributor. This allows two way conversation or links to be set up without allowing feedback to occur between the two systems or studios
Clip i	Audio has exceeded the maximum levels of the system and has moved from linear representation to distortion
Clip ii	A small segment from a longer audio item
Clock i	Physical device to show the time
Clock ii	A timing control signal between two or more devices or systems to ensure synchronization between those systems
Codec	Coder-Decoder, device which encodes audio from an analogue or digital format into a compressed data stream and will also decode a return compressed signal back to analogue or digital audio

Compression i	Method of reducing the dynamic range of an audio signal to create a most consistent level
Compression ii	Method of reducing the data rate of an audio signal for transmission or storage in an electronic data system, Mpeg is a common example
Compressor	Device which performs compression on an audio signal to control the dynamic range of that signal
Control Room	A technical area which controls the audio output from a studio. Often used by producers and technical operators running a studio based programme.
CPU	Central Processing Unit, the core of a computer that performs all of the work
CR	Control Room
Cross Talk	Where signals from one electronic system unexpectedly interfere with another, causing interference
CRT	Cathode Ray Tube, old style computer or TV screen now being replaced by TFT, LCD or Plasma
CTA	Central Technical Area, area where all of the main technical equipment that links all studios and transmitters together.
Cue i	An indication to a presenter or producer that a programme element is ready, which may be a light on a machine indicating it is ready to play, or a lamp or audio indication that a presenter should start talking
Cue ii	An audio communication circuit from an outside source to a studio that is used for production purposes, to pass information relating to a programme element
DA	Distribution Amplifier, used to distribute one audio source to many destinations
DAB	Digital Audio Broadcasting.
DAT	Digital Audio Tape, playback and record device using DAT tapes, fully digital
dB	Decibel, a logarithmic measurement scale, of audio and other items
DDI	Direct Dial In, a public phone number which will be delivered to an internal phone system
DEL	Direct Exchange Line, phone line with public number delivered to a single handset and not part of an internal phone system
Denon	Manufacturer of professional audio equipment
Destination	Place where an audio signal is being sent to
Digigram	Manufacturer of professional computer sound cards for both AES and Analogue audio
Distortion	Where audio ceases to be reproduced faithfully by an item of equipment, possibly due to a fault, more commonly due to audio levels exceeding the designed maximum

DLS	Dynamic Label Segment, DAB service used to carry regularly changing text to a DAB receiver, often used for news headlines, travel updates and information about the song playing now.
DOS	Disk Operating System, computer operating system from Microsoft before windows came along
Dubbing	Duplication or copying of audio from one method of storage to another, such as CD to CD or CD to DAT
E1	Large data circuit between two sites, with the correct equipment attached could carry computer data, audio or video
Earth	Connection to known zero volts potential, same as ground
Echo	A system where a sound source is repeated often more than once, possibly whilst new sounds are being added
EDAC	Multi-pole audio connector, used for many audio connections on a multi-core cable
Engineer	Person who has a qualification in the field of engineering, including, mechanical, chemical, electronic etc…
EPG	Electronic Programme Guide, contains details of the programme you are listening to, or future programmes yet to happen
Expander	Device which controls audio levels within user set criteria
Fax	Facsimile, device which scans paper documents, transmits them via telephone lines, and are reproduced on paper at a remote destination
Feedback	Any system where the input is fed by the output of the same system, commonly noticed in the audio domain as squeals, whistles or rumbles where a microphone is too close to the speakers it is feeding
File	Collection of related data, held in one place on a computer system
Finalizer	A brand of audio processor used to compress, change or modify the sounds applied to it.
Floppy Disc	Removable data storage device on a computer
FM	Frequency Modulation, method of encoding audio for radio transmission
Format i	The style of a radio stations programmes, such as speech, news or music
Format ii	A pre-stored mode or state of an electronic system, such as the operation mode of a studio mixer
F-Series	Screw connector for high frequency signals, such as radio frequency
Fuse	Electrical safety device to shut down power supply in the event of a fault, such as excessive current being drawn
Fused	The state of a device which has been shut down by a fuse to protect it from a fault
FX	Effects, sounds or processes applied to an audio signal to change the original audio

G722	Model of ISDN Codec suitable for low quality speech.
Gate	A method of turning off an audio source which is too quiet, as defined by user settings
Generator	Method of producing electricity, which can power buildings or studios or outdoor events, usually run on petrol or diesel
Gentner	Brand of studio phone in and TBU systems
Glensound	Make of studio and field based audio Codecs, suited to voice
Ground	Connection to known zero volts potential, same as earth
Headphones	Miniature or medium sized loud speakers worn on head
Headroom	Safety zone within electronic systems, often referring to maximum space between the "normal" audio levels and distortion due to excessive level
Hessian	Fabric often used as final decorative finish to an acoustically treated wall
Hub	Computer device connecting several machines together, mostly superseded by switches and routers
I	Term referring to electrical current (I)
IEC	Mains voltage power connector, also known as kettle plug
Internet	World wide network of computers allowing data to be passed between computers, companies and systems
Intranet	Internal version of Internet, allows information to be shared internally between staff
ISDN	Integrated Services Digital Network, a common digital telephone line, allowing a fixed rate of data to be connected on demand between two devices, often used in radio for connecting remote locations to deliver audio using ISDN Codecs
Jackfield	Flexible audio patching system allowing a normal audio connection between say CD player and Mixing Desk to be changed at short notice to deliver different audio
KiloStream	Data circuit provided by telecoms companies to permanently connect a site with a remote site at a fixed data rate, based on kilo bits per second data transfer
Krone	Flexible audio connection system, allowing items of equipment to be connected together for long term use, although can be easily changed
Krone Block	Multi-pole balanced audio connector designed to be mounted upon a frame as part of an installation, allows fixed installation items to be connected together and also patched or re-routed where necessary
Krone Cork	Device to be inserted into a krone block to temporarily stop or break an audio path
KVM	Keyboard Video Mouse
KVM Switcher	Device which will allow multiple computers to be connected to one KVM, or many KVM to be connected to one computer

Lamp	Light bulb, either for illuminating a room, or providing an indication
LED	Light Emitting Diode , miniature lamp, ideal for indication
Limiter	Device used to control audio levels that exceed a user defined maximum, often used to prevent distortion
Lindy	A brand of KVM extender
Linear	Term used to describe non compressed digital audio
Log	A list of events, either created as things happen so they can be interrogated later, or a list of events that need to happen in the future, such as a play-list which a computer play out system will follow
Long Wave	Transmission Band, used for broadcast
Loud Speaker	Device for converting electrical signals into audio
LS	Loud Speaker
Mains Hum	Where mains frequency unexpectedly appears on a signal such as an audio feed, often caused by poor grounding or shielding
Mast	Construction, pole or tower for mounting antenna upon
Master Clock	Timing device which distributes time information to other devices, such as studio clocks or synchronization information between electronic systems. The master clock is often controlled by an external timing reference such as GPS
Master Control(MC)	Brand of computer software used to play audio in studios
Match-pack	Device for converting unbalanced audio to balanced, and vice versa
Matrix	A routing system, usually for audio that allows multiple audio sources to be made available to multiple destinations, controlled by users remotely from the matrix, often the back bone behind outside source switching
MC	Master Control
MCR	Often confused with a CTA, although an MCR has more of an operation role, where programmes can be controlled in the MCR in terms of manually setting levels, selecting between studios or including additional audio
MDU	Mains Distribution Unit
Medium Wave	Transmission Band, used for broadcast
MegaStream	Data circuit provided by telecoms companies to permanently connect a site with a remote site at a fixed data rate, based on mega bits per second data transfer
Meter	Visual indication of a measured quantity, in audio terms the meter will reflect the audio level at the measured point
MIC	Microphone, converts sound to an electronic signal which can be recorder or mixed with other audio sources
Mic-arm	A flexible sprung stand for a microphone to be mounted upon

Minidisk	Digital audio recording and playback medium, similar to CD only smaller and can be recorded. Uses a light digital compression to allow more audio to be stored on the small disk
Mix Minus	Poorly performing method of achieving a clean feed. The outside source audio is subtracted from the main programme mix in order to create a pseudo clean feed, does not perform as well as a clean feed due to differences in level and propagation delays in the system
Mixer	Mixes multiple audio sources together, such as microphones, CD, DAT etc, to produce an output. Studio mixers have controls to allow the operator to set audio levels in the mix
Mix-pack	Small stand alone mixer normally used to mix two or three audio signals together without any control over level. Used as a temporary fix rather than a permanent installation
Monitor i	Loud Speaker
Monitor ii	Computer display screen
Mono	Single channel of audio, not stereo
MP2	A digital compression type to reduce the data rate of an audio feed. Stored computer audio files using MP2 compression, would be known an MP2 file
MP3	A digital compression type to reduce the data rate of an audio feed. Stored computer audio files using MP3 compression, would be known an MP3 file
MPEG	A digital compression type to reduce the data rate of an audio feed, or size of stored computer audio file. Both MP2 and MP3 are built on the MPEG concept and standards
MSF	Standard of distributing time code by radio waves to clocks
Multiplex i	Method of encoding, multiple digital audio signals into one digital stream than can be recovered and separated at a remote end
Multiplex ii	A DAB multiplex broadcasts a suit of stations and services within a single data stream, a multiplex must be received before individual services can be decoded by the listener
Music Line	A high quality audio link provided by telecoms company using a light low loss digital compression
Music Taxi	Brand of ISDN Codec for audio links, using MPEG coding
Neumann	Brand of high quality microphones for voice work within studios
Normalise	A process of setting standard levels to audio files
N-Type	Screw connector for high frequency signals, such as radio frequency
OBIT	Obituary, an alarm from a news organization that significant public figure has died, or a pre-recorded programme or feature about a public figure who has or may die
Omnia	Brand of high quality audio transmission processors
Online	Refers to a system or computer that is connected and working

Orban Optimod	Brand of high quality audio transmission processors
OS	See Outside Source
Outside Source	An audio feed from outside of the studio or station environment that has been made available within the station, a source of audio from outside
Patch Field	Flexible audio connection system, designed in its normal state to connect audio sources and destination as per the station design. The normal audio routes and paths can be changed easily by the insertion of Patch cables
Path	The route taken by a signal from one place to another
PC	Personal Computer
PCB	Printed Circuit Board, base upon which electronic components are based to build circuits upon
PCM	Pulse Code Modulation, a method of representing analogue signals, such as audio in a digital format
PF	See Pre-Fade
Phone Op	Person who operates telephones, often with a studio environment
Phonebox	Brand of studio phone in system, using computer display screens to capture and share caller information, and route those calls to the mixer for broadcast
POTs	Plain Old Telephone line
Powerlog	Brand of audio logging system for both legal and production playback purposes
PPM	Peak Programme Meter, displays audio levels for the user based around perceived audio levels rather than actual measured, very good method of monitoring audio levels
Pre-Fade	Method of listening to an audio source on a mixer without that audio source going to air
Presenter	Person who presents radio programmes
Prima	Brand of MPEG ISDN audio codec
Printer	Device which converts computer based documents or data into paper
Producer	Person who manages radio programmes and radio presenters, may also plan and research the programme content
Profanity Delay	Device which delays station output by several seconds to allow profanity by presenter, guest or callers to be deleted before transmission on live programmes
Pronto	Brand of ISDN Codecs using the Mpeg standard
ProTools	Multi-track audio editing system allowing many sounds to be played in sequence or at the same time, with effects and edits to create a new audio item
PSU	Power Supply Unit, device which provides and regulates power to an electronic device

Rack	A vertical cabinet into which rack mountable equipment may be housed, normally 19inches wide
Rack Mount	An item of technical equipment which can be mounted by design into a 19inch rack of equipment
Racks	See CTA
Radio Car	Mobile studio which via radio or satellite links allows live broadcast quality audio to be generated and fed back to the station for live or recorded broadcast
Radio Text	An information service built into an FM radio transmission that can provide details such as, song, weather, news and travel information to be displayed on an FM receiver at the same time as listening to the station
RCD Trip	Residual Current Device, electrical safety device to shut down power supply in the event of a fault
RCS	Radio Computer Services, company who provide computer systems for music scheduling, play-out, reconciliation and traffic management
RDS	Radio Data Services, an information service built into an FM radio transmission that can provide details such as, song, weather, news and travel information to be displayed on an FM receiver at the same time as listening to the station
Receiver	Receives radio signals
Reverb	Device for creating decaying echoes on an audio signal
Revox	Manufacturer of professional audio equipment, best known for quarter inch tape machines
RF	Radio Frequency
Router i	Device which routes audio in a configurable way, could be used as part of an outside source system, similar to a matrix
Router ii	Device which connects computer systems together and routes data from one machine to another without allowing another machine to see that data
RTB	Return Talkback (incoming talkback)
Sadie	Multi-track audio editing system allowing many sounds to be played in sequence or at the same time, with effects and edits to create a new audio item
Short Wave	Transmission Band, used for broadcast
Shortcut Editor	Brand of stand alone (not networked) audio editing machine, ideally suited to quickly editing interviews or phone calls
Simulcast	Programme that simultaneously goes out on more than one transmission service
Sky	European satellite broadcaster
SMS	Short Message Service, allows short text messages to be sent between mobile phones
Sonifex	Brand of professional audio equipment

Sony	Brand of professional and domestic audio equipment
Soundscape	Multi-track audio editing system allowing many sounds to be played in sequence or at the same time, with effects and edits to create a new audio item
Source	A place or device generating audio
Spatial Enhancer	Device which widens a stereo audio image to make the source sound as if it is coming from outside of the speakers rather than within them
Split	A common programme going to many areas may contain extra local information, which is known as a split area, where main programme is suspended to allow local information to be inserted
Spot	A commercial
Spot Set	A group of commercials
Stereo	Audio signal containing two separate audio channels, left and right, providing a spatial image to the audio
Studio Manager	Person who manages the audio in a studio and may control the studio mixer and any audio sources and levels, can also be known as a Technical Operator
Switch i	Device to turn something on or off
Switch ii	Device which connects computers together and routes data between them
Switch iii	Device which connects telephones together and routes calls internally or externally
Sync	Synchronization, signal or connections between systems to ensure they operate at the same time
TA	Traffic Announcement, signal sent to FM RDS radio receivers informing the radio or listener that Traffic and Travel information is being broadcast
Tag Block	Method of providing electrical or electronic connection
Talk Back	Communication channel between studios or outside sources that allows a non broadcast conversation to take place to aid programme generation or production
Tandberg	Brand of satellite receiver
TBU	Telephone Balance Unit, device which converts telephone audio standards to studio audio standards and allows phone calls to be recorded or broadcast
TDM	Time Division Multiplexing, digital standard for communication or switching systems which allowing multiple signals to be combined onto one circuit and recovered remotely
Tech-Op	See Studio Manager
Telemetry	Method of interrogating a system or device to find out its operational status, often used in fault finding, may automatically report any problems found
TFT	Transistor Field Technology used in flat screen computer monitors

Threshold	User set audio level which causes something to occur, such as compression, expansion or limiting.
Track	A channel of audio in the context or recording or play back
Traffic	Commercials, or department within a radio station that looks after commercial scheduling
Transmitter	Device which radiates energy for the purpose of broadcasting, with the intention of it being received and listened too
Tuner	Device which "tunes" in to a specific radio signal and allows it to be listened too
Turntable	Record Player
UHF	Ultra High Frequency, transmission band used for broadcast
UPS	Un-Interruptible Power Supply
VDU	Visual Display Unit, electronic screen displaying information, usually a computer or TV screen
VHF	Very High Frequency, transmission band used for broadcast
Videologic	Brand of DAB Digital Radio Receivers
Voice Track	A computer recorded track of a presenter linking two items of audio together, to be played in the future by a computer automation system
WAV	Wave File, a computer stored audio file, usually linear
Wharton	Brand of synchronized time systems and studio clocks
Word Clock	A time reference signal between audio systems to ensure multiple audio devices are synchronized
XLR	3 pole balanced audio connector, provides ground, phase signal and anti-phase signal

Permissions & Thanks

This book has been made possible by the good people I have worked with at some great radio stations over the last 15 years. Many of whom have simply tolerated me as I learned my trade. Some of whom would still say I have plenty to learn, and I would not deny that. The radio stations include, Northern Air Hospital Radio, University Radio Hull, Classic FM, Lite AM, Wish FM, Heart 106.2, Heart Digital and LBC. Plus many other stations that for one reason or another touched my life. Thanks must also go to Chrysalis Radio who gave permission for me to include the odd photo here and there of different parts of the infrastructure and installations.

Some of the best people I have ever worked with and have personally contributed to my knowledge include the technical folks at Chrysalis Radio, Classic FM, Clyde Broadcast Products and RCS. Hopefully once published they won't immediately denounce me as an idiot. The London Engineering Beer Club will be the ultimate jury on the matter.

Commercial Radio is about making money, however, the passion for radio is what allows the money to be made by providing the best tools to the best people to make the best radio. If I have played a useful role in achieving that then I am satisfied. If this book is purchased by anyone who goes on to achieve great things then I am more than satisfied.

Thanks must go to you the reader for parting with your cash, and hopefully reading to this point.

My wife and family have all been very kind in tolerating me during the writing of this book. Their contributions have included moral support, proof reading, offering advice and the odd meal here and there.

Finally in the true friendly style of radio it would be wasted opportunity and an insult to the medium to not say a quick hello to everyone else who knows me.

Lightning Source UK Ltd.
Milton Keynes UK
UKOW012114071012

200189UK00001B/8/A